ROADSIDE GEOLOGY OF THE
YELLOWSTONE COUNTRY

ROADSIDE GEOLOGY
OF THE
YELLOWSTONE COUNTRY

by
William J. Fritz

Roadside Geology Series
MOUNTAIN PRESS PUBLISHING COMPANY
MISSOULA, MONTANA

Fifth Printing, July 1992

Library of Congress Cataloging-in-Publication Data

Fritz, William J.
    Roadside geology of the Yellowstone country.

    Bibliography: p.
    1. Geology—Yellowstone National Park Region—Guide-
books. 2. Yellowstone National Park. I. Title.
QE 79.F75   1985   557.87'52   85-4934
ISBN 0-87842-170-X

Mountain Press Publishing Company
P.O. Box 2399
Missoula, MT 59806
(406) 728-1900

"I am convinced that, at its best, science is simple—that the simplest arrangement of facts that sets forth the truth best deserves the term science. So the geology I plead for is that which states facts in plain words—in language understood by the many rather than by the few."

—George Otis Smith, 1921,
Director, U.S. Geological Survey 1907-1930.

MAP SYMBOLS

MAP SYMBOLS	AGE	ERA	PERIOD	EPOCH	Representative formation Names of rocks exposed in the Yellowstone Cou[ntry]
	.01	Cenozoic	Quaternary	Recent	Hot springs, landslides, streaм glacial, lake deposits
Recent, glacial, stream gravel lake silts, hot spring deposits	1.6			Pleistocene	Rhyolite, welded ash, basalt о & 3rd volcanic cycles
Landslides	5		Tertiary	Pliocene	Huckleberry Ridge Tuff of 1st volcanic cycle
Basalt flows	24			Miocene	Six Mile Creek Formation
Rhyolite welded-ash	27			Oligocene	Renova Formation
	58			Eocene	Absaroka Volcanic Supergrou Willwood Formation
Rhyolite flows	66			Paleocene	Fort Union Formation
Tertiary valley fill		Mesozoic	Cretaceous		Everts Formation Eagle sandstone, Telegraph Creek Formation, Cody Shale Various sands and shales near Cody
Tertiary basalt and andesite flows	144		Jurassic		Morrison Formation Ellis Group
Eocene volcanic gravel (Absaroka Volcanic Supergroup)	208		Triassic		Gypsum Springs Formation Chugwater Formation
Intrusive rock	245				
	286	Paleozoic	Permian		Phosphoria Formation
Mesozoic sediments	320		Pennsylvanian		Tensleep/Quadrant Sandstone Amsden Formation
			Mississippian		Madison Group
Paleozoic and Mesozoic sediments	360				Three Forks Shale Jefferson Dolomite
	408		Devonian		No rocks in Yellowstone Country
			Silurian		Bighorn Dolomite
	438		Ordovician		Snowy Range Formation Pilgrim Limestone Park Shale Meagher Limestone Wolsey Shale Flathead Sandstone
Paleozoic sediments	505		Cambrian		
	570				
Precambrian basement all rock types	4700	Precambrian			Granite, gneiss, schist, amphibolite gabbrо dikes, pegmetite of crystalline basement

vi

Preface

I wrote this book to introduce the non-geologist to some of the interesting history written in the rocks of the Yellowstone Country. I hope to help residents and visitors better appreciate the countryside, and to provide a more meaningful trip to some out-of-the-ordinary areas. This book is not for my professional geological colleagues. It assumes no previous knowledge of geology or science. However, I hope that some of my colleagues will find this a useful guide for their first visit to the area, or for leading students on field trips.

It was difficult to decide in what direction to write the road-guides because not everyone will drive in the same direction; thus either choice would disappoint some. I settled on a rule to write the guides from north to south, and from west to east. I did depart from this general scheme in a few places because that is the way I usually drive that particular stretch of road.

Because this is a general book that touches on many aspects of the local geology, I have summarized the published work of a myriad of gelogists. I hope that I have done justice to their interpretations and I thank them for their work. In the back of this book, I list some of the more readily available references on the local geology. Because this book is intended for the general reader, the list is far from complete for a research scientist, and does not acknowledge all outside papers that I read. If you are really interested in researching the geology of the Yellowstone Country, most of the technical literature is cited in the biblio-graphies of the general references. My own research in the

Yellowstone Country has involved work on both the paleobotany and the depositional environment of the fossil forests. For this reason, I tend to concentrate on the detail of geology in the northern area of Yellowstone Park—not because the geology to the south is any less interesting, but because I am more familiar with that to the north. In addition to my own research, I have led geology students on field trips over most of the roads covered in the roadguide section. I thank these students for their perceptive questions that forced me to learn even more of the geology.

I compiled the geology on the roadguide maps from several sources. Especially helpful were the following maps: U.S. Geological Survey, 1972, Geological Map of Yellowstone National Park, Map I-711 at 1:125,000; Love, J. D. and Christiansen, A. C. (compilors), 1983, Preliminary Geologic Map of Wyoming at 1:500,000; Ross, C. P., Andrews, D. A., and Witkind, I. J., 1955, Geologic Map of Montana at 1:500,000; and reports by Pierce (1979) and Montagne and Chadwick (1982 WGA Guidebook). I have generalized the geology from these technical maps so that the non-geologist reader may more easily interpret the general geologic history without getting bogged down in details. There are two important generalizations you should keep in mind when using the roadguide maps. 1.) Rather than differentiating individual formations, many similar units of rock have been combined. Examples include all of the Quaternary volcanics, combined into two or three units, and all Mesozoic or Paleozoic rocks, combined into one unit. 2) Most of the thin skim of recent stream, lake, glacial, and soil cover is excluded. Thus the map units refer to underlying bedrock; overlying deposits are shown only when very thick, extensive, and are obscuring all bedrock.

To get the most out of the book I suggest that you first read the introductory chapters, and then read the entire roadguide for your planned trip. As you drive, keep track of the geology on the roadguide maps. If you carefully read the entire book you will find some repeat of information and descriptions—this is intentional. I tried to make each roadguide as self-contained as possible, so that it can be used without a lot of searching for a previous description of something. It thus, by necessity, repeats information in places. A glossary of technical terms is included at the end of the book.

I thank Dave Alt for his inspiration and help in writing this book. I also owe a great deal to Lanny H. Fisk who first introduced me to the Yellowstone Country. I thank the staff of Yellowstone National Park for their cooperation both in my past research in Yellowstone and for reviewing this book; however, the park is not responsible for errors contained in this volume. I especially appreciate the personal friendship and support of this project extended by Tim Manns, north district naturalist and park historian. Lanny H. Fisk, Loma Linda University; Wayne Hamilton, Physical Science Coordinator, Yellowstone National Park; Roderick A. Hutchinson, Park Geologist, Yellowstone National Park; Timothy E. La Tour, Georgia State University; Robert L. Christiansen, U.S. Geological Survey; Norman A. Bishop; Jana Morman; Jenifer Hutchinson; Sylvia Harrison; Richard M. Ritland; and Timothy R. Manns, North District Naturalist and Park Historian, Yellowstone National Park; and other staff of Yellowstone National Park provided valuable suggestions and critical reviews of the manuscript; I am grateful for their time in reading several drafts and in giving innumerable helpful suggestions to improve both content and style. Frank Drago, Georgia State University, provided welcome drafting assistance with several maps in the introductory chapters.

Bill Fritz
Georgia State University
Atlanta, Georgia
October 1984

CYCLE	VOLCANIC UNIT	AGE (millions of years)
	Plateau Rhyolite	
	Central Plateau Member	0.07-0.2
	(forms Pitchstone, Madison, and and Solfatara plateaus)	
	Mallard Lake Member	0.15
	Shoshone Lake Tuff	0.18
	* Obsidian Creek Member	0.09-0.32
	* Roaring Mountain Member	0.08-0.4
Third	(makes Obsidian Cliff)	
Volcanic	Upper Basin Member	0.28-0.6
Cycle	Osprey Basalt	0.2
	Madison River Basalt	0.1 -0.6
	Swan Lake Flat Basalt	0.2 -0.6
	(flow at Sheepeater Cliff)	
	Falls River Basalt	0.2 -0.6
	LAVA CREEK TUFF	**0.6**
	Undine Falls Basalt	0.7
	Mount Jackson Rhyolite	0.8
Second	Island Park Rhyolite	1.3
Volcanic	**MESA FALLS TUFF**	**1.3**
Cycle	Basalts of Warm River	0.6 -1.2
	Lewis Canyon Rhyolite	0.9
First	Basalt of the Narrows	1.5
Volcanic	**HUCKLEBERRY RIDGE TUFF**	**2.0**
Cycle	Junction Butte Basalt	2.2

* Lava flows outside the caldera

Stratigraphic list of selected volcanic units of the Yellowstone Plateau volcanic field. Each unit may contain from one to numerous flows. Boldface welded-ash tuff units punctuate the explosive climax of the caldera-forming phase of each volcanic cycle. Note that each cycle starts with pre-caldera lava flows, climaxes with an explosive caldera-forming rhyolitic welded-ash, and continues to ooze rhyolite and basalt lava flows in a post-caldera phase. Compiled from Christiansen and Blank (1972) and Hildreth and others (1984).

Table of Contents

Grand Geyser in winter. National Park Service Photograph.

I
Introduction

Rocks exposed in the Yellowstone Country provide a record of past events covering a little less than two-thirds of the entire history of the earth. The earth is slightly older than 4.6 billion years. The geologic story of the Yellowstone Country begins with rocks at least 2.7 billion years old that are mashed remnants of even older pre-existing rocks and continues with processes that still shape the land's surface. In the next few pages I will attempt to summarize this history, trying to hit the highlights and concentrate on features you can actually see if you follow the roadguides. Three of the more interesting features are the fossil forests, glaciation, and Quaternary volcanism. The individual roadguides point out the specifics.

I wrote this book for people with no training in geology. Nevertheless, I use some technical words. Definitions appear in the glossary at the end of the book.

Get out of your car to see the rocks. If you want to collect specimens, visit localities outside of Yellowstone National Park, because collecting is prohibited in the park. All rocks exposed in the park also occur on roads outside.

When hiking in the park, it is always advisable to check with a park ranger for weather conditions, level of streams to be crossed, bear sightings, etc. Also, stay out of all hot springs and thermal features, and remain on the boardwalks and paths. Outside the park, I suggest several side trips on dirt roads. Check before taking any of these because snow and impassable mud holes can remain well into the late spring and early summer.

The glacially gouged valley of Rock Creek at the eastern edge of the Beartooth Plateau. Photo by Dave Alt.

II
Overview of
Yellowstone Country Geology

The oldest rocks in the Yellowstone Country are 2.7 billion-year-old Precambrian basement rock: plutonic rocks that cooled from molten magma at great depth, and metamorphic rocks that formed as igneous and sedimentary rocks were heated and squeezed at great temperatures and pressures miles below the earth's surface. Most of the plutonic rocks are pink granites, which contain large crystals up to a quarter of an inch across. Granite contains four different types of mineral grains. Take a few minutes to stop at one of the granite outcrops mentioned in the roadguide and see if you can identify these types. Clear glassy quartz grains make up about 30% of the rock. Two different kinds of feldspar comprise the remaining 60-70%. Plagioclase feldspar is chalky white and rich in sodium and calcium. Pink grains are potassium feldspar which contain potassium instead of sodium. Most of the granites of the Yellowstone Country are pink because of all the potassium feldspar. The remaining dark flecks in the granite are micas, amphiboles, and various other black minerals that contain abundant iron and very little silica. The dark minerals are collectively called mafic minerals. Some granites, the granitic gneisses, contain mineral grains that were stretched and lined up when they were mashed during mountain building.

Metamorphic rocks form as other rocks are heated and squeezed deep within the crust of the earth. Even though rocks seem very solid, they flow like putty when buried and heated,

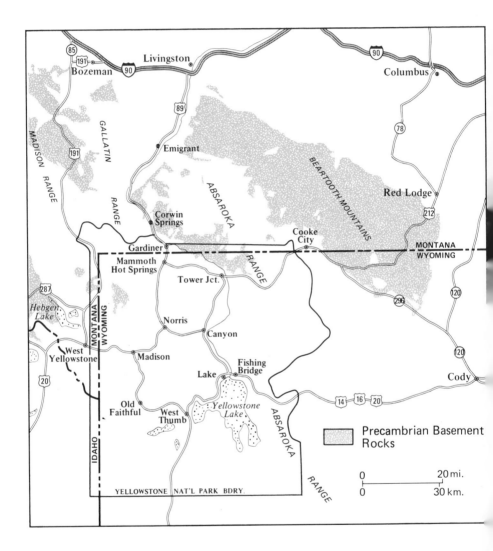

Location map of roads covered in this book showing major exposures of Precambrian crystalline basement rock.

but not quite melted. Heat and pressure cause new minerals to grow and large streaks to form in the rocks. Metamorphic rocks with thick streaky bands are called gneiss (pronounced "nice"), and rocks that split into flakes because they contain many mica grains oriented parallel to each other are schists. The oldest rocks in the Yellowstone Country are 2.7 billion-year-old granitic gneisses. However, even these ancient rocks must have had ancestors.

After formation of the granites and gneisses, a second period of rock formation occurred. During this time a dark magma very low in silica injected cracks in the older rocks. This rock can be seen as dark thin diabase dikes cutting at odd angles through the pale granite.

One of the best areas to see Precambrian basement rocks is on the Beartooth Plateau (see the Cooke City to Red Lodge roadguide) where they have been uplifted and all of the younger rock cover stripped off. These basement rocks are also exposed in the center of many of the large folds (anticlinal arches) in the area, including the Gallatin River canyon south of Bozeman, Yankee Jim Canyon north of Gardiner, Rattlesnake Mountain anticline west of Cody, and in the Lamar Canyon in northeast Yellowstone Park.

After the early Precambrian igneous and metamorphic rocks formed and hardened, there was apparently a long period of uplift and erosion in the Yellowstone region that brought them to the surface. Sediments eroded from the basement rocks made late Precambrian sedimentary rocks of the Belt Supergroup, now exposed throughout western Montana and northern Idaho. In the area covered in this guide, buried erosional surfaces (geologists call them unconformities) are all that remain as a record of this time interval. After the erosional period, Paleozoic and Mesozoic sedimentay rocks were deposited in shallow seas, swamps, rivers, and lakes. Many of these formations contain fossils of many different types of plants and animals now extinct. Some of the better collecting localities outside of Yellowstone National Park are mentioned in the appropriate roadguides. Because the philosophy for the existence of national parks includes preservation of all natural and historic objects for future visitors to enjoy, no collecting is allowed inside the park boundaries. Photographs, however, are always appropriate!

After deposition of the early Mesozoic sedimentary rocks, a period of uplift and mountain building called the Laramide orogeny occurred between 100 and 50 million years ago as North America collided with the Pacific oceanic crust and at least two small continents that now form parts of the Pacific Northwest. Such collisions happen because the surface of the earth is a series of plates that slowly move. New oceanic crust forms at spreading centers along mid-ocean ridges that exist in most ocean basins, like the mid-Atlantic Ridge. Old ocean crust sinks where two plates collide. Collision of North America and the Pacific ocean floor with several small continents shoved the Rocky Mountain region up into a long linear mountain chain. Forces tilting the area caused Paleozoic rocks around Cooke City to break loose and slide over younger rocks east of Cody (see the 212-296 junction to Cody roadguide).

Soon after the uplift, about 50 million years ago, volcanoes started to erupt in central Idaho, southwestern Montana, Yellowstone and northwestern Wyoming to deposit the Absaroka Volcanic Supergroup, a pile of andesite lava flows, basalt flows, airfall ash, mudflow conglomerate, stream gravel and sand. In places, this debris buried spectacular forests now petrified and exposed in southwestern Montana and northwestern Wyoming. The Absaroka volcanoes stopped erupting about 40 million years ago. Following this period of volcanism, only uplift and erosion occurred in the park until volcanism started again about 2.5 million years ago in a region called the Yellowstone Plateau volcanic field. Also, some stream and lake deposited sediments exist in the valleys south of Bozeman and Livingston. In some places, these recent sediments are fairly thick. Lake clays, silts, and sand in Hayden Valley and some of the travertine rocks around Mammoth Hot Springs are over 200 feet thick.

Following a long period of erosion, volcanoes again erupted in the Yellowstone Country to produce the Yellowstone Plateau volcanic field. This time, a series of three large volcanic eruptive cycles climaxed in the Yellowstone Country between 2 million and 600 thousand years ago. Intermittent smaller eruptions, including the one that made a small caldera now filled by the West Thumb of Yellowstone Lake less than 150,000 years ago, continued until about 70,000 years ago. These eruptions shaped part of the Yellowstone Plateau, filling a very large caldera or crater-like basin. Hot magma at depth still heats water for all the hot springs and geysers. After the last eruption, two periods of glaciation helped shape the surface. Streams are still eroding, depositing, and continuing to alter the geology of the Yellowstone Country.

Even today, the Yellowstone Country is very active geologically. Uplift that may have started this century continues today at almost an inch per year along a football-shaped area in the central part of Yellowstone and could be the prelude to future volcanic eruptions. This uplift and other mountain building forces throughout the area make earthquakes that shape the land and, in Yellowstone, alter some of the thermal and geyser activity. For example, the Mount Borah earthquake located in central Idaho in late 1983 slightly increased the average time between the eruptions of Old Faithful and caused previously dormant springs to erupt again as geysers. Similar and even more dramatic events were produced by the 1959 Hebgen Lake earthquake located in southwestern Montana just northwest of West Yellowstone.

Yellowstone Lake National Park Service Photograph.

7

Petrified redwood tree buried by gravel of the Sepulcher Formation just west of Tower Junction. Before the fence was built collectors got every scrap of several other large stumps nearby.

8

III
Fossil Forests

Ancient sandstone and gravel deposited along the flanks of an old chain of Eocene (50 million years old) volcanoes bury some of the finest deposits of petrified wood anywhere. Certainly, many areas in the United States contain petrified wood; many even have some upright stumps. None of these other areas, though, contains so many upright trees, such well preserved wood, and such good exposures as the petrified forests in Yellowstone.

Petrified wood in the Yellowstone Country was buried by mudflows and sediment-laden streams during eruptions of two chains of volcanoes in Eocene time, about 50 million years ago. Geologists call these volcanic sediments and associated lava flows the Absaroka Volcanic Supergroup. Two concentrations of petrified wood exist in these rocks. The Lamar River Formation contains stumps and trees in the northeast corner of Yellowstone Park and extends into Montana and Wyoming around Cooke City. The Sepulcher Formation, part of the Washburn Group, contains the petrified trees in northwest Yellowstone and southwestern Montana. Both formations were deposited about the same time, but from eruptions of different volcanic peaks.

Both the Lamar River and Sepulcher formations accumulated-on an erosional landscape full of hills and valleys. Both formations rest on an Eocene erosional surface (an unconformity where some of the story is missing) carved on Paleozoic

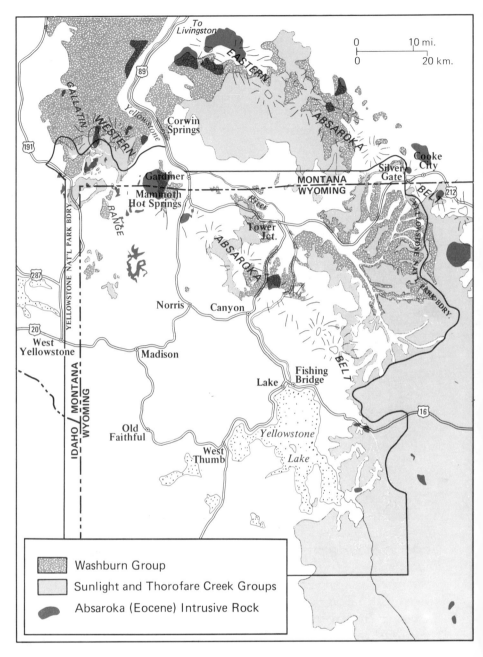

Map of Eocene volcanic gravel and lava flows (Absaroka Volcanic Supergroup) in the Yellowstone Country. The intrusive rock occurs at the sites of the ancient volcanoes that made the two parallel Eocene mountain chains, here called the Eastern and Western Absaroka Belts. The Sepulcher Formation of the Washburn Group buries "fossil forests" in northwestern Yellowstone and the Lamar River Formation of the Washburn group preserves the petrified trees in the northeastern part of the park. Compiled from Chadwick (1970), Smedes and Prostka (1972), USGS (1972), and Fritz (1982).

marine sediments and Precambrian granite and gneiss. The volcanoes continued to erupt after deposition of these two formations, covering them with other formations in the Absaroka Volcanic Supergroup. Some of these other formations, especially in the vicinity of Sylvan Pass, contain a few scattered buried stumps and logs, but not as many as in the Lamar River and Sepulcher formations.

Three upright trees petrified where they grew. These can be seen on a hike to near the top of Specimen Ridge.

Visitors to the Yellowstone Country have been fascinated by the petrified trees ever since Jim Bridger and other mountain men of the mid-1800s told of the Yellowstone forests of stone. However, long before Bridger's visit, Native Americans told legends involving the same forests. Early explorers made large collections of the wood and other plant fossils between 1872 and the turn of the century.

Besides the petrified wood, numerous other plant fossils exist in the sediments, including leaves (some of which still have the cuticle or organic tissue preserved), needles, cones, and pollen grains. Although small, pollen grains are almost indestructible, and easily fossilized. They help identify the different kinds of trees, shrubs, and plants that grew in the forests. Almost 200 curiously different types of plants are in the fossil forests.

Trees such as spruce, fir, alder, willow, and elm that today grow best in cool climates were buried with petrified trees usually associated with warm sub-tropical climates, such as breadfruit, magnolia, podocarpus, aralia, avocado, laurel, and possibly a type of mangrove. Besides plants from these two seemingly climatic extremes, a host of intermediate fossil plants from warm places, such as hardwood forests in the eastern United States also exist. Some of these are sycamores, walnut, dogwood, oak, maple, and hickory. While remarkable diversity approaching that of the fossil forests exists in the modern forests of South China, the southeastern United States, and parts of Central America, often the zonation caused by a few thousand feet of elevation contributes to this diversity. Some transport of pollen, leaves, and even stumps and logs from the sides of the volcanoes may help to account for the mixed assemblage of fossil species. Because all these different trees cannot grow at the same elevation, those that prefer cool climates may have been washed down the mountain to mix with their warm temperate or tropical neighbors.

Imagine the temperate forms, like those whose descendants

Typical stream gravel conglomerate burying fossil trees in the Lamar River Formation.

still live in the Yellowstone Country, growing at high elevation on the volcanic mountain ranges while the tropical to warm temperate plants thrived in the warm valleys. Similar zonation of tropical lowlands to cool-temperate highlands occurs in volcanic mountainous regions today. Guatemala is a good example. There, most representatives from the fossil forests live in small geographic areas with a ten thousand foot change in elevation. Similar conditions must have existed in Yellowstone during Eocene time.

There are a number of good places to see the fossil forests when driving through the Yellowstone area. Most are mentioned in the following roadguide, but I will also describe some of the better locations here. Some of the best known localities that the early explorers studied occur in the Lamar and Soda Butte valleys, between Tower Junction and Cooke City. Most of the dark brown clffs along this route contain petrified trees in the Lamar River Formation. Few maintained trails lead to the cliffs, so be ready for a rugged off-trail hike. One of the easiest hikes to some of the best trees is on Specimen Ridge south of Slough Creek to the three tall vertical stumps shown in several park publications. However, the rocks burying the trees are not well exposed in the grassy slopes.

To see the rocks, hike up Amethyst Mountain from the Lamar River picnic area. Wade the Lamar River and proceed directly up the slopes in the distance. Numerous stumps and logs and excellent rock exposures appear throughout the cliffs. Before you climb Amethyst Mountain, it is wise to check with a local ranger for bear closures and to be sure the Lamar River is low enough to wade safely. Some years the river remains dangerous well into June, even early July. Hiking boots suitable for rugged off-trail use, a day pack, and a warm sweater are always advisable.

Other good exposures exist on the flanks of Mount Norris, reached from the Cache Creek trail head, and on the cliffs of Mount Hornaday, behind the Pebble Creek campground. Both of these cliffs are treacherous and full of loose rocks, so walk with extreme care. A hard hat should be worn at all times because the loose rocks can roll with very little disturbance; many are even knocked loose by curious mountain sheep. Also, all natural features in Yellowstone Park, including petrified wood, are protected against collecting. Even a small amount of collecting can ruin an area for future visitors and can damage the locality for scientific research.

Fossil forests in the Sepulcher Formation occur north of Gardiner in the Tom Miner Basin, and north of West Yellowstone along Specimen Creek. Because the Tom Miner Basin is outside of the park, it is one place where you can collect small bits of petrified wood with a permit from the U.S. Forest Service office in Gardiner. The Petrified Tree just west of Tower Junction is also buried in rocks of the Sepulcher Formation.

Close inspection of the rocks burying the forests reveals their history—and these rocks surely tell a fascinating story. Rocks that bury the fossil forests came from two chains of volanoes bordering a long valley. The volcanic rocks belong to four main types: Gravel conglomerate made of pea- to washtub-size chunks of hardened lava; ash sandstone layers made out of water-deposited sand-size ash grains; airfall ash deposits; and a few lava flows. Most of the volcanic material, both reworked sediments and original lava flows, are andesites, a lava intermediate in composition between black low-silica basalt and pale high-silica rhyolite. Andesite comes in various shades of brown, green, or purple. A little bit of basalt is found in the Sepulcher and Lamar River formations.

When the volcanoes erupted, they caused massive landslides and rock debris to cascade into streams coming off the sides of the mountains. The debris mixed with water from the streams and from melting snow, making a soupy sediment flow that rushed down the streams—remember photographs of the Toutle River right after Mount St. Helens erupted. These mud and sediment flows deposited a chaotic mixture of mud, sand, and gravel. Near the peaks, mudflow conglomerate layers are tens of feet thick and contain cobbles as big as washtubs. As the flows moved, they picked up rocks from the stream bed, and

A collection of transported trees and trees buried where they grew by the 1980 Mount St. Helens mudflows along the Toutle River. This is identical to the environments that buried trees in the Eocene fossil forests.

Petrified logs all lined up by streams and buried in volcanic gravel in the Lamar River Formation.

mixed them with broken volcanic rock and uprooted trees. Farther away from the mountain, the flows become thinner and their rock fragments smaller until they pass into thin layers of gravel, sand, and mud in the valleys. Eruptions and mudflows did not destroy entire forests. Instead, the mudflows came down stream valleys so their deposits are only as wide as the stream bed, ranging from a hundred feet to a mile or more wide.

After mudflows choked the streams, water reworked the rock debris into stream conglomerate and sandstone. The flowing water carried the mud particles away and left well-sorted sand or gravel deposits with cross-beds and other evidence of stream deposition. Some mudflows dammed the streams to form small temporary lakes where silt and clay settled out of suspension into thin horizontal shale layers interbedded with coarse-grained gravel from the mudflows and streams. Many of the best preserved leaf fossils occur in lake shales.

15

Fine-grained shales of mud and ash deposited in lakes dammed by mudflows. These shales contain the best fossil leaves.

In addition to sedimentary rocks, some of the Lamar River and Sepulcher formations are made of true hot volcanic rocks. On Amethyst Mountain, there is a thick white layer visible from the road about one third of the way up the slope, a blanket of airfall ash that was probably not reworked by streams. Very near the volcano, on rocks exposed along Cache Creek, this layer was hot enough to weld itself together, and for new crystals to grow after it was deposited. Near the base of the Lamar River Formation there are four basalt flows; one makes up a low ridge in the meadow below Amethyst Mountain. The best basalt flows are at the base of Mount Hornaday, a short hike from Pebble Creek campground. The top of the flows are scoriaceous, full of cavities called vesicles formed by gas bubbles that did not escape from the lava before it hardened. All lava contains dissolved gas that is kept in solution by high pressure. As the pressure is reduced, the gas expands to make cavities which are most abundant at the top of the flow. An example of a similar phenomenon occurs when you take the cap off a bottle of soda and the carbon dioxide gas makes bubbles. Near the top of Amethyst Mountain, a flow of andesite lava came down an old stream channel. The base of the flow is pillowed from contact with the water in the river.

A small lake formed when Mount St. Helens mudflows dammed a small stream killing the trees. Lake shales in the fossil forests were deposited in a similar manner.

Stratified mudflow deposit on the North Fork of the Toutle River near Mount St. Helens. Similar deposits bury the fossil forests.

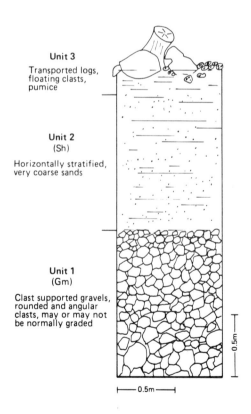

Unit 3
Transported logs, floating clasts, pumice

Unit 2
(Sh)
Horizontally stratified, very coarse sands

Unit 1
(Gm)
Clast supported gravels, rounded and angular clasts, may or may not be normally graded

0.5m

├── 0.5m ──┤

A diagram that interprets the three distinct rock types in the mudflow shown in the preceding photo. This sequence occurs throughout the fossil forests.

17

The petrified trees themselves, in addition to the rocks, also tell the story of the fossil forests. Not all of the petrified trees are tall vertical stumps shown in most of the articles written about the fossil forests; many exist as horizontal logs. These are sometimes difficult to see because they get buried by the loose talus rolling down the slope—you must look closely to find them. Near the old vents, nearly all of the wood exists as horizontal logs while exposures away from the volcanoes have 20-60% horizontal logs and 40-80% upright stumps. Most logs were lined up in one direction by the streams into log jams.

A graph that shows that there are fewer upright stumps near the old volcanic peaks. Also the grain size of the sediments is much finer grained away from the volcanoes.

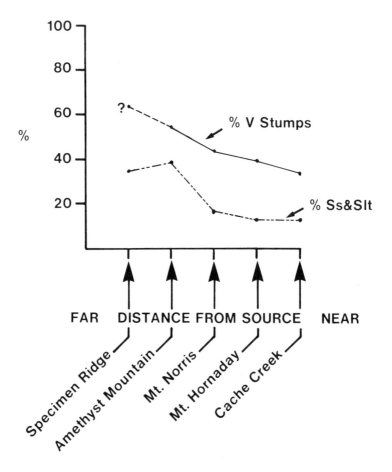

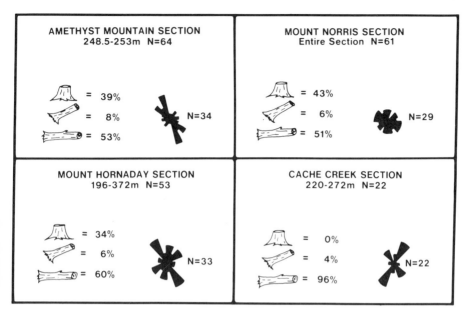

Representative percentages of logs and stumps on various exposures of the Lamar River Formation. The "rose" diagrams show how the logs are all lined up by streams and mudflows.

In addition to the tall upright trees that were buried where they grew, a few stumps are short, because they were snapped off by the eruptions or by mudflows. These stumps were probably transported by the streams and buried in gravel some distance from where they grew. Unlike the tall stumps, the short ones seldom have fossil soils associated with the roots. Out in the valleys, however, soils occur at the base of many tall trees. A good example of one fossil soil exists at the base of the three tall upright trees buried where they grew and now exposed on Specimen Ridge. The soil can be recognized by a band of lighter colored tan or yellow rock near the roots.

Many of the fossil trees do not have bark because it was eroded off before they were buried or because it decayed away. Rare redwoods, however, do have some of the bark preserved. A small number of the logs in the fossil forests are leaning at a steep diagonal angle. Presumably they were pushed over by the flows. One on Amethyst Mountain is nearly 30 feet long.

The petrified trees in Yellowstone are so well preserved that with only a hand lens you can often separate them in the field to one of three basic types. The most common basic type is red-

19

A petrified stump with bark on Mount Hornaday.

wood, which has thin growth rings made up of only one kind of cell. The growth rings are caused by seasons; the tree grows fast in the spring and summer and stops in the fall and winter, producing a growth ring. Trees of any of the three types that grow in the tropics grow all year long and have no rings. A second type of tree is pine, which has a few scattered large cells called resin ducts that store the pitchy sap. The third kind are broad-leaved trees, which have many larger cells called vessels that help move water up the trunk to the leaves. These trees also have strong rays that radiate from the center of the trunk. Many of the broad-leaved trees have no growth rings thus confirming the mixture of plants that grew in different climates. Using this description and the accompanying photos you should be able to identify some of the better preserved trunks.

The trees are so well preserved because all of the organic tissue or wood is still present. If you were to place a chip of the petrified wood into an acid that dissolves the minerals (hydrofluoric or hydrochloric acid), brittle wood cells would still remain after the acid had done its work. Wood contains many hollow tubes that move water to the leaves and needles. After the wood is buried, ground water fills the tubes and plugs them with silica and calcite dissolved from the volcanic ash. This happens in the same way that a leaking faucet becomes clogged with scale in areas with hard water. This type of petrification

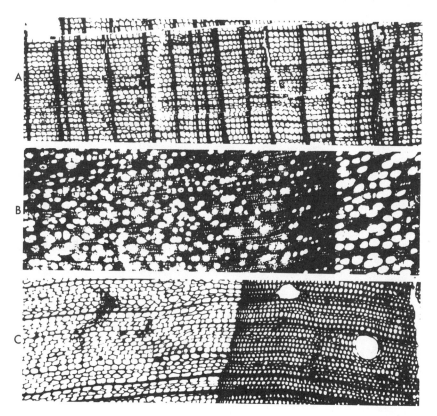

Photos of growth rings of three petrified stumps taken through a microscope and magnified about 45 times. A is a redwood, B a tropical leaf tree, and C a pine.

allows the preservation of minute detail. Markings on the cell wall only a few microns across (the size of a red blood cell) visible only at a thousand power under a microscope are still preserved on these 50 million-year-old trees. This is a very different kind of petrification than is common in Petrified Forest National Monument in Arizona. There, Jurassic trees have been completely replaced with silica and only a few have remaining organic wood.

Most of the petrified trees in Yellowstone are full of silica, a small number, of calcite. A very few have some cells filled with calcite and others with silica. When these sit at the surface, acidic rain water dissolves the calcite but not the silica, leaving small holes that look like insect borings. Some of the trees started to rot before they were petrified, and have hollow centers. If you are lucky, you may see one in which large purple

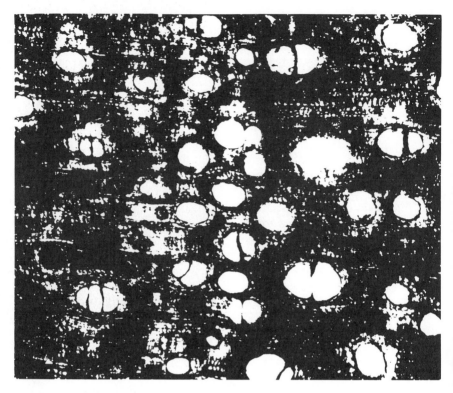

A view of a petrified tropical hardwood tree magnified about 50 times. The large white ovals are vessels that carry water to the leaves.

amethyst quartz crystals fill the cavity. Amethyst Mountain was named for such crystals that were numerous when the first explorers visited. Unfortunately, many have since been removed by collectors. If you find some, please take only photos so future visitors may enjoy the crystals, and so you won't have to pay a fine!

Some predictable changes occur in the sedimentary rocks that bury the fossil forests between the highlands near volcanic peaks and the lowland valleys that separated the two mountain chains. Near the peaks, all conglomerate layers are very coarse-grained and contain boulders the size of washtubs and cars. Some sorted stream gravel separates these debris flow units, but very few sand lenses can be found. Most petrified trunks are horizontal logs all lined up in one direction where they were pushed over by debris flows. Most of the wood in these high energy sections belongs to the pine family—trees that could have grown at high elevation where it remained cool

Coarse-grained boulder gravel from a mudflow on Mount Hornaday. This looks identical to mudflows on the Toutle River deposited after the eruption of Mount St. Helens. The biggest rocks are about the size of a bushel basket.

throughout the year. These areas also contain lava flows, sills, dikes, and deposits of welded-ash flows from clouds of hot ash that stuck together forming solid rock. Rocks deposited in what was once a valley 50 million years ago, like those that are now exposed on Amethyst Mountain, Specimen Ridge, and Specimen Creek, are much finer grained. They contain less conglomerate and the fragments are mostly the size of baseballs. Many more lenses of stream deposited ash-sandstone occur out in valley sections.

Both stream conglomerate and sandstone show evidence of water reworking of mudflow debris. Numerous features formed by flowing water such as channel cut-and-fill sequences, trough cross beds, ripple cross laminations, graded beds, and load casts can be found. Some of the fine-grained lake silt and clay occurs interbedded with the stream deposits.

A much higher percentage of upright petrified trees exists away from the ancient peaks. In the center of the valley, they comprise 40-80% of the wood. Because mudflows in modern environments, such as Mount St. Helens, only deposit up to 10% of the transported wood as upright stumps, most of these petrified trees must be preserved where they grew. However, identification of the stumps and logs here shows a mixture of tropical trees and ones from the cool uplands. A few of the upright stumps, as well as many of the horizontal logs, were transported in streams and mudflows to produce the mixture.

The 1980 explosion of Mount St. Helens provided a modern region to observe processes similar to those that buried the fossil forests. The largest blast on the morning of May 18, 1980, destroyed a large area north of the peak leaving logs lined up like match sticks, overturned heavy equipment, and devastated forests. However, these features occur on ridge tops where they will erode away before they are buried and preserved in the geologic record. Lowland stream valleys where mud was deposited are the best places to compare to the fossil forests.

After both the big eruption and a smaller one in 1982, great mudflows poured down the sides of Mount St. Helens into the north and south forks of the Toutle River, Muddy Creek, and Cowlitz River and, after flowing over 75 miles, into the Columbia River. The deposits left by the mudflows are identical to ones burying the fossil forests in the Sepulcher and Lamar River formations in the Yellowstone Country. The mudflows produced an obvious vertical sequence that you should be able to recognize in the Eocene fossil forests. The base of the unit contains grapefruit-size cobbles that get smaller upward to pea-size gravel. This layer is overlain by horizontally layered sand. The top unit is a mixture of huge boulders plus transported stumps and logs. The unit varies from six to ten feet thick near the mountain to only a few inches downstream at the Columbia River. The largest cobbles exist near the

Mudflows on the sides of Mount St. Helens similar to ones that rushed down the sides of the Eocene volcanoes and buried the fossil forests.

Cliffs on Amethyst Mountain. Dark layers that are not continuous are stream gravels. Notice the numerous petrified stumps.

mountain and then the size of the sand and gravel gets smaller downstream as in the conglomerate layers in Yellowstone Park.

When I visited the Mount St. Helens area shortly after the eruption, it was just like Yellowstone! I found many horizontal logs all lined up by the streams and mudflows and some upright stumps that had been moved by the flows propped upon the stubs of their roots. I found that about 10 percent of the transported trees remain as upright stumps, the rest as horizontal logs. The mudflows also buried many standing trees where they grew along the edges of stream channels. Thus, in Yellowstone when you find concentrations of over 10 percent upright stumps, some were preserved where they grew alongside stream channels. A few million years from now when the Mount St. Helens sediments have hardened into rock and the trees have petrified they will be almost like those in the Yellowstone Country. Both the mudflows and the appearance of the trees is identical.

There are two important respects in which the Mount St. Helens model is less than perfect, however: considerable logging accounts for part, but not all, of the stumps and logs and the side blast caused more breakage than many eruptions. Nevertheless, the phenomena of transport is beautifully exhibited and much of the transported remains at Mount St. Helens is from source areas not disturbed by man.

One additional difference between Mount St. Helens and the fossil forests is that in Yellowstone there were many volcanoes erupting, intead of a single eruption. Remember that during Eocene time two chains of volcanoes existed in the Yellowstone Country with a long narrow valley between them. The floor of this valley must have been less than a thousand feet above sea level to allow the tropical trees to flourish. The peaks were from five to ten thousand feet higher than the valleys, and remained cool enough to allow cool climate plants like spruce and fir to grow. As the volcanoes erupted, they dumped debris into the basin from both sides. Some trees were uprooted and transported from the highlands and deposited on the sides of channels with lowland vegetation that was not moved—thus causing the mixture of plants. After the trees were buried the cells were plugged with silica and calcite to petrify them before they could rot.

As you visit the fossil forests, try to observe the features that I describe in this chapter. To see all the features, visit several localities to compare rocks that were deposited near the peaks to ones further out in the valleys. I have tried to indicate the localities closest to the road in the appropriate roadguides; however, most require a hike of some sort.

Petrified pine stump with well-preserved growth rings on Amethyst Mountain.

IV
Quaternary Volcanism

Yellowstone is probably best known for its hot springs and geysers. More than 350 thermal features in the park have erupted as geysers and about 200 to 250 of these spout each year. Some erupt like clockwork every few minutes, others gush sporadically, and may go for years between eruptions. Steamboat Geyser, in the Norris Geyser Basin, is the highest recently active geyser in the world. Over 100, or nearly a fifth of all of the geysers in the world, are in a square mile around Old Faithful—reason enough for a visit. Only about 200 active geysers occur outside of Yellowstone National Park scattered in 30 basins throughout the world. All of Yellowstone's steaming wonders (there are over 10,000 geysers, hot springs, fumaroles, mud pots, etc. in the park) exist because hot rock still cools only a mile or two beneath the surface of the Yellowstone Plateau. The history of the cooling magma and the past volcanism makes a story as fascinating as the geysers themselves.

All of central Yellowstone National Park is a caldera, a giant collapsed crater of a not-so-extinct volcanic field. The history of Quaternary explosive volcanism and thermal features in the Yellowstone Country starts roughly two million years ago with a tremendous volcanic explosion that made a large caldera older than the Yellowstone caldera and produced over 5000 times the volume of volcanic material of Mount St. Helens. Volcanism has continued intermittently since before the first

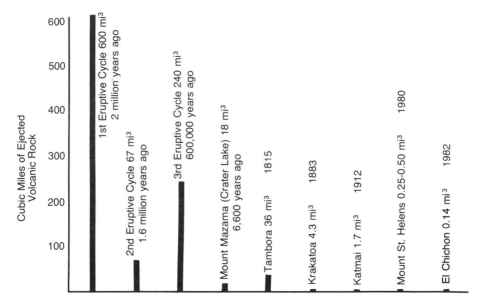

Graph showing the amount of volcanic welded-ash ejected during the explosive caldera-forming phases of the Yellowstone Plateau volcanic field, as compared to major volcanic eruptions during historic times. Volumes calculated from Smith and Braile (1982 WGA Guidebook).

explosion until at least 70 thousand years ago. However, the volcanic activity was punctuated by two additional major eruptive climaxes at about 1.3 million years ago (produced the Mesa Falls Tuff from the Henrys Fork caldera) and 600 thousand years ago (produced the Lava Creek Tuff from the Yellowstone caldera). Each of these eruptive climaxes was part of a volcanic cycle that included smaller explosions and lava flows.

No one knows exactly what causes volcanism in Yellowstone or elsewhere. Perhaps it exists over a place where a plume of hot rock rises from deep within the mantle. Most continental volcanoes erupt along the edges of continents in chains parallel to ocean trenches, where the dense basaltic ocean floor sinks into the earth's interior under the light granite of the continental crust. As the ocean basalt sinks it becomes hot and melts.

28

Like wax in a lava lamp, the hot magma is less dense than cooler solid rock, so it rises to the surface and makes a volcano a few miles inland from the sea. Mount St. Helens and all of the Cascades formed in just such a fashion. But Yellowstone, sitting alone about 700 miles inland, is something of a puzzle, and makes the mantle plume idea attractive to some geologists.

To explain the mantle plume, think of what happens deep in the earth's mantle where the rocks are very hot, probably partly molten. Part of the heat was caused by radioactive decay, and some may even remain from the time the earth formed. These very hot rocks tend to behave like a liquid, and, like any heated liquid, produce convection currents. Think of heating a kettle of thick pea soup, how the plumes of soup rise before it boils. The mantle may contain a similar plume of rising hot rock that begins to melt as it nears the surface. If this plume exists under a continent it melts its way to the surface and makes a giant volcano. Eventually such a hot spot can weaken the continental crust and split the continent in two. This has probably happened many times in the past such as when Africa and the Americas were ripped apart. Yellowstone and the nearby Snake River Plain of Idaho with Craters of the Moon may represent the start of such a split.

Whatever the cause, volcanic activity began about 2 million years ago. The first climactic eruption in the Yellowstone Plateau volcanic field produced over 600 cubic miles of mostly rhyolitic magma—the extrusive equivalent of granite—which makes a very dangerous volcano. Molten rhyolite is extremely viscous, and it may reach the surface heavily charged with gas. If that happens, the magma will erupt in a violent explosion of superheated steam and other gases.

The first climactic eruption occurred 2 million years ago as a complex of caldera-forming eruptions from three overlapping areas that made a large composite hole in central Yellowstone National Park. As the magma erupted, the resulting hollow collapsed forming a depression called a caldera that later filled with viscous rhyolite lava that was too stiff to flow great distances. The only trace that remains of the rim of this caldera lies beyond the southern border of the park and west into and around Island Park, Idaho. However, most of the rhyolite traveled great distances all over the Yellowstone Country by a different method. The explosion shattered much of the magma

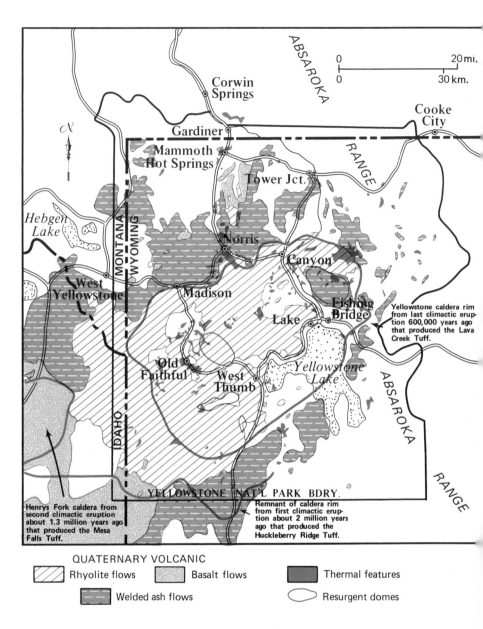

Map showing caldera rims from the three major eruptive cycles over the past 2 million years in the Yellowstone Plateau volcanic field, and the various volcanic rock types produced by those eruptions. Also shown are the two resurgent domes and concentrations of thermal features. The volcanic rocks are covered in places by a thin skim of recent glacial and stream gravel not shown on this map. Compiled and redrafted from information in Christiansen and Blank (1972), USGS (1972), Keefer (1971), Smith and Christiansen (1980), and Smith and Braile (1982).

into fine dust that flowed as a dense cloud suspended in red hot gases. When the particles came to rest, they were still hot enough to weld themselves together into a dense rock. These flat lying particulate (or "pyroclastic") flows, called the Huckleberry Ridge Tuff, cover much of the Yellowstone Country, and lie both on and beneath lava flows of both rhyolite and basalt. Basaltic magma is very fluid so it can produce lavas that flow a long way.

There were at least three of these big rhyolite eruptive cycles in the Yellowstone Country that make the Yellowstone Plateau volcanic field. Each started with pre-caldera lava that oozed to the surface followed by a large climactic explosion and welded-ash flows that drained the magma chamber, causing a collapse and formation of a large caldera. Finally, in the post-caldera phase, more lava oozed out to partially fill the caldera. By the explosive caldera-forming part of this cycle, more than 800 cubic miles of volcanic welded-ash alone, in addition to lava flows, was produced by the three large events. By comparison, this is nearly ten thousand times the amount of explosive debris erupted during the 1980 exlosion of Mount St. Helens!

Diagram illustrating how a welded-ash unit forms.

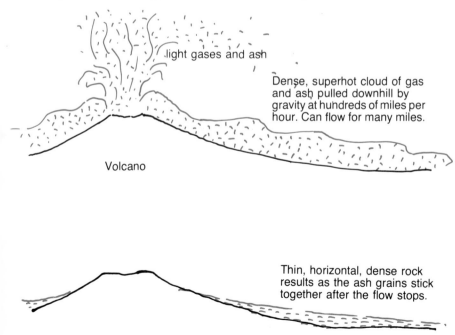

light gases and ash

Dense, superhot cloud of gas and ash pulled downhill by gravity at hundreds of miles per hour. Can flow for many miles.

Volcano

Thin, horizontal, dense rock results as the ash grains stick together after the flow stops.

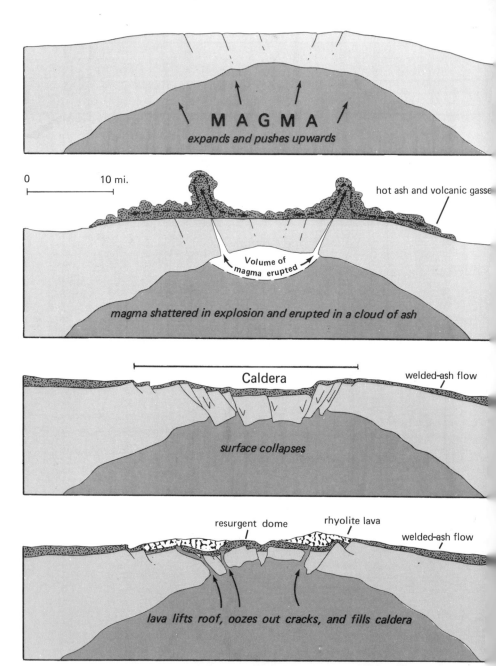

MAGMA
expands and pushes upwards

0   10 mi.

hot ash and volcanic gasses

Volume of
magma erupted

magma shattered in explosion and erupted in a cloud of ash

Caldera

welded-ash flow

surface collapses

resurgent dome   rhyolite lava

welded-ash flow

lava lifts roof, oozes out cracks, and fills caldera

Diagram illustrating the sequence that made the Yellowstone caldera during the third eruptive cycle 600,000 years ago. Similar events marked the first two cycles as well. Note that resurgent doming occurred early in the last stage. Post-caldera lava eruptions and further uplift occurred slowly over a long time period. Redrafted and modified from information in Keefer (1971) based on Christiansen and Blank.

32

The second eruptive cycle was a relatively small one (still one of earth's great eruptions!) that climaxed around 1.3 million years ago just outside Yellowstone National Park in the Island Park area of Idaho. This climactic eruption from the Henrys Fork caldera (nestled in the southwest corner of the 2 million-year-old caldera) produced 67 cubic miles of welded-ash. Most of these pyroclastic flows, called the Mesa Falls Tuff, are outside the park in Idaho and none exist along the main roads in the park.

The third eruptive cycle was climaxed by explosive events from two overlapping areas that created the Yellowstone caldera of central Yellowstone National Park that is over 47 miles long and 28 miles wide. Much of this third caldera rim can still be seen as a series of low hills ringing the central plateau. However, in places the rim has been covered by subsequent lava flows. This last major series of eruptions, 600 thousand years ago, produced about 240 cubic miles of welded-ash and basalt flows that cover much of Yellowstone National Park. Because fewer younger rocks cover the lava from this third cycle, geologists can tell a more complete story of the history of this cycle. Apparently lava oozed periodically from the central part of the Yellowstone Plateau volcanic field for nearly half a million years after the end of the second cycle 1.3 million years ago. This pre-caldera oozing ended with the violent caldera-forming series of eruptions 600 thousand years ago. During this time the thick welded-ash flows came so fast that many individual flows all later cooled as one unit.

Since the explosive, caldera-forming eruption of the third cycle, there have been intermittent post-caldera eruptions along fissures and a few explosions from 600,000 up to about 70,000 years ago. One of these smaller eruptions, slightly less than 150,000 years ago, left a secondary caldera that is now West Thumb of Yellowstone Lake. Also between 80 and 70 thousand years ago the youngest rhyolite lava flows that now make the Pitchstone Plateau and other flows were extruded. Since these last flows there has been no lava produced; however, that could change in the future. A large football-shaped area between Old Faithful, Fishing Bridge, and Canyon is uplifting and swelling at nearly an inch per year. If this swelling continues, more eruptions might occur.

The hot springs and geysers show that the Yellowstone volcano is probably still active because they result from the cooling of the hot rocks and magma that still exist below the surface of the caldera. Even though very hot, much of the rock beneath the caldera may now be solidified, partly because the high pressure that exists several miles below the earth's surface keeps it solid. However, earthquake waves and shock waves from man-made explosions show that some partially molten rock may still exist under the northeast edge of the caldera and elsewhere. Molten rock transmits waves much differently than solid rock. This partial melt probably has an oatmeal-like consistency with solid crystals floating in a viscous liquid. If the pressure over the area were to be suddenly reduced, possibly by the continued expansion of the resurgent domes or by earthquakes, lavas could erupt or gases in the partial melt might explode in another great volcanic eruption.

Hot springs and geysers form when rain and snow melt seep into the earth and come in contact with the hot rock. Hot water is less dense than cold water, so it rises toward the surface. If it is warmer than the average annual air temperature and flows freely to the surface, a hot spring, warm spring or hot pool results. If, on the other hand, the path to the surface is plugged, possibly by sinter, a winding fracture system to the surface or a vertical column of water, the weight of the water keeps that at the bottom from boiling. Its temperature may rise as high as 500 degrees Fahrenheit or hotter before it boils. Finally, the tremendous pressure from the hot water trying to expand clears a path to the surface and some water leaks out, causing the pressure to drop. The remaining water then flashes into steam that blows the entire column of hot water and steam out as a geyser. This extremely high temperature of some of the water shows that hot volcanic rock is only a mile or two below the surface. As you walk among the picturesque pools and enjoy all of the thermal features throughout the Yellowstone Country, think of the series of events that caused their formation and of the hot rock cooling beneath you.

V
Glaciation

During the last 2 million years, glaciers covered much of the Yellowstone Country on at least two different occasions; probably many more. They left their record in the scenic landscapes. The earliest widely preserved glacial time or stage was sometime between 160 and 130 thousand years ago. For convenience, geologists call this the Bull Lake glacial event. The more recent Pinedale glaciers formed between 70 and 13 thousand years ago and left the most obvious features. Later glaciers destroyed much of the evidence of the earlier ones.

It is difficult to tell just why these two glacial events happened when they did, but it must have involved a global climate change, because glacial deposits of these ages occur widely in North America and Europe. Most people think of glaciers forming in extremely cold regions—but this is not always the case. In the Olympic Mountains of western Washington, active glaciers flow to within about 3,000 feet above sea level. Farther inland, permanent snow fields and glaciers can hardly form in the Wind River Range at 12,000 feet above sea level. The difference is precipitation. More snow falls each winter in the Olympic Mountains than can melt during the summer. Wyoming is so dry that not enough snow falls to make a glacier. What is needed to make a glacier is not cold weather, but more snow in the winter than melts in the summer. Eventually the snow will pile up to a depth where it is squeezed into ice from pressure caused by the weight of the

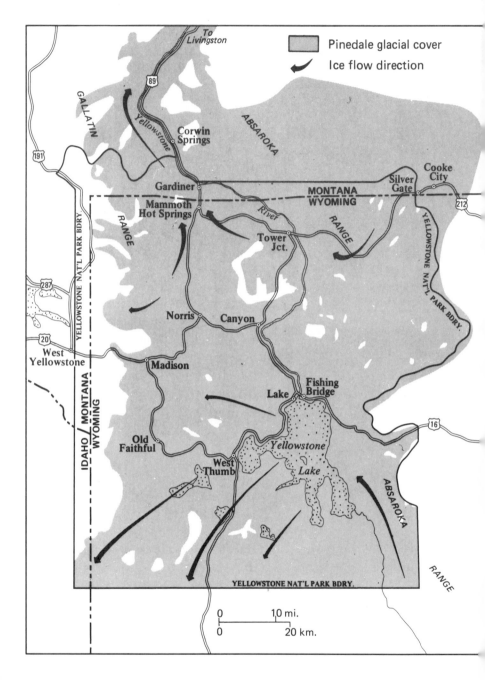

Map showing the area covered by glacial ice during Pinedale glaciation about 13 thousand years ago. From Keefer (1971) and Pierce (1979).

overlying snow and starts to flow as a glacier.

The glacial deposits in the Yellowstone Country must represent a time of cool wet climate—possibly brought on by all of the ash and sulfate particles released into the atmosphere during an intense period of volcanic activity. Volcanoes can produce enough ash, gas, and sulfate particles in the air to cause a slight cooling—maybe even enough to start glaciation. The 1982 eruption of the volcano El Chichón in southern Mexico released so much sulphur into the atmosphere that some scientists blamed this for the cool summer the following year. Whatever the cause, glacially scoured landforms, as well as sediment deposited from glaciers, occur throughout the Yellowstone Country.

Mountain glaciers—those most common in this area today—start when snow accumulates to around 200 feet deep, is squeezed into ice, and starts to flow down the valley. As it moves, the glacier plucks bits of rock and carves the mountains at the head of its valley into a steep-walled amphitheater called a cirque. Where glaciers erode all sides of the mountain, a gnarled pinnacle called a horn peak remains. In the Beartooth Mountains, northeast of Yellowstone, Pilot, Index, and the Bear's Tooth peaks are good examples of glacial horn peaks. Sometimes this erosion leaves a long thin knife edge ridge,

The landscape on top of the Beartooth Plateau consists essentially of deep canyons cut into a broad upland surface. Dave Alt Photo. 37

called an arete, between two horn peaks. As the glaciers flow down old stream valleys, the ice erodes the valley into a steep-walled glacial trough. All of the eroded rock fragments plucked and embedded in the ice act like sandpaper to scrape and polish the bedrock surface. Many glacial grooves and striations exist on granite in the Beartooth Plateau.

Glaciers also left many depositional landforms in the Yellowstone Country. Glaciers carry along a lot of rock debris plucked from the bedrock. Other debris comes from talus falling onto the glacier from the walls of the valley. Most glaciers have long ridges of this sediment riding along their edges. When the glaciers melt back, the ridges of sediment remain as moraines made out of till—a poorly sorted sediment that contains all grain sizes from silt and clay to huge, car-size boulders. When the glaciers retreat, the ice continues to flow forward, but the glacier melts at its toe faster than the ice advances. Heavy snow fall or cool summers make the glacier advance because the ice moves faster than it can melt at the toe. If the snow is light or the weather warm, the glacier retreats because ice melts faster at the toe than it advances. In either case, the glacier continues to flow.

Where the glacier melts, all of the rock debris in the ice drops into a pile of till called a terminal or end moraine that marks how far down the valley the ice came. If it had flowed further it would have destroyed and reworked the first moraine. The largest terminal moraines indicate the glacier remained in one place for a long time; a large conveyor belt that dropped till to build the terminal moraine.

As the glacier retreats, large boulders drop to form glacial erratics. There is a whole field of them east of Tower Junction on the road to Cooke City. Sometimes, glacial ice becomes buried in the till. When the ice finally melts, the ground collapses to make a depression that fills with water. Because till is poorly sorted, water cannot seep into the ground, and fills the depression to form a kettle pond. Melt water carries away some of the till and deposits it as stream sand and gravel. The steep-walled valleys carved by the glacier are often unstable without the support of ice. For example, sand can only be made into a 30-40 degree slope before it collapses, unless it is held up by a retaining wall. Where glaciers cut steep walls in weak rocks, large landslides move after the ice melts. Some steep valley walls stand until an earthquake jolts them into motion. Landslides have a hummocky, poorly drained surface with aspen groves and springs. The toe is a steep wall 10 or 15 feet high and is easy to see; such as those on the northeast slope of Sepulcher Mountain seen from Gardiner. It is also possible to look back up on the wall and see the scarp cliff where the landslide started.

Look as you drive through the Yellowstone Country for these glacial landscapes along many of the roads in the following roadguides. Some exist beside the road, others are visible in distant mountain ranges. All are numerous and easy to identify. If you use these descriptions and photographs you should become an expert in identifying the work of ancient glaciers.

Landslide that flowed into the Lamar Valley after glacial ice carved a steep wall and then melted away. Notice the scarp, hummocks, and steep toe with an aspen grove.

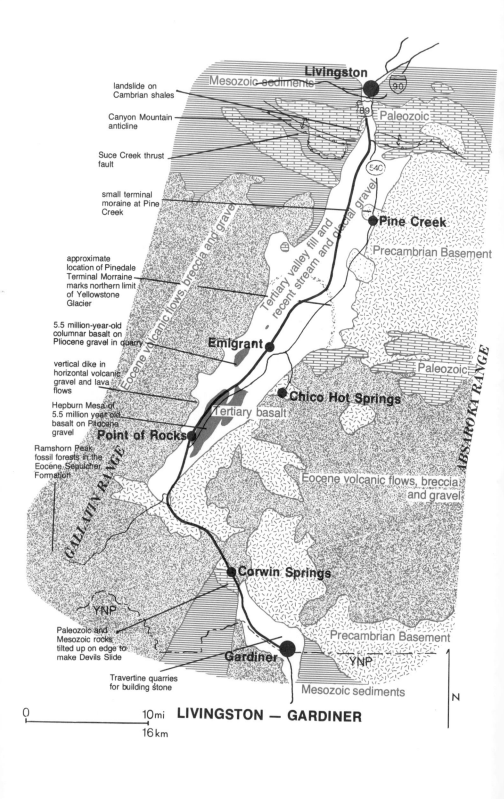

landslide on
Cambrian shales

Canyon Mountain
anticline

Suce Creek thrust
fault

small terminal
moraine at Pine
Creek

approximate
location of Pinedale
Terminal Morraine
marks northern limit
of Yellowstone
Glacier

5.5 million-year-old
columnar basalt on
Pliocene gravel in quarry

vertical dike in
horizontal volcanic
gravel and lava
flows

Hepburn Mesa of
5.5 million year old
basalt on Pliocene
gravel

Ramshorn Peak,
fossil forests in the
Eocene Sepulcher
Formation

Paleozoic and
Mesozoic rocks
tilted up on edge to
make Devils Slide

Travertine quarries
for building stone

Livingston

Mesozoic sediments

Paleozoic

Pine Creek

Precambrian Basement

Tertiary valley fill and
recent stream and glacial gravel

Emigrant

Paleozoic

Chico Hot Springs

Eocene volcanic flows, breccia and gravel

Tertiary basalt

Point of Rocks

ABSAROKA RANGE

GALLATIN RANGE

Eocene volcanic flows, breccia
and gravel

YNP

Corwin Springs

Precambrian Basement

Gardiner

YNP

Mesozoic sediments

N

0   10mi   **LIVINGSTON — GARDINER**
     16 km

VI
ROADGUIDES

Livingston—Gardiner
50 mi./81 km.

The town of Livingston is built on a thin skim of stream gravel brought from the south by the Yellowstone River. The recent stream gravels near town cover late Cretaceous to early Tertiary sediments made of volcanic sand and gravel.

Livingston is at the northern end of the narrow Paradise Valley, at the south end of a large valley defined by the northern end of the Gallatin range on the west and Absaroka range on the east. Just south of town, U.S. 89 enters a narrow canyon cut by the Yellowstone River through uplifted Paleozoic and Mesozoic sedimentary rocks. At the entrance of the canyon, the rocks dip north, and then fold sharply and dip south as the road leaves the canyon. This fold is the Canyon Mountain anticline. Paleozoic rocks were shoved to the south over Mesozoic rocks on the Suce Creek fault, causing buckling and folding. A large landslide in slick Cambrian shales blocked the road near the center of the anticline in 1967 where a roadcut undercut the shale.

From the canyon south, the road follows the valley of Yellowstone River called "Paradise Valley" by many residents. This valley owes its form to several large glaciers that flowed down it from mountains to the south, leaving outwash and moraines. Moraines are composed of glacial till, a chaotic mixture of silt, sand, pebbles and large boulders dumped directly from the glacial ice. When the ice melted, the debris remained behind in large windrows that record the stages of glacial retreat. The moraine surface is very hummocky in linear ridges that cross the valley to mark the end of the glacier, and follow the edge of the valley to mark the sides of the glacier. By contrast, outwash is well sorted sand and gravel deposited from streams of

41

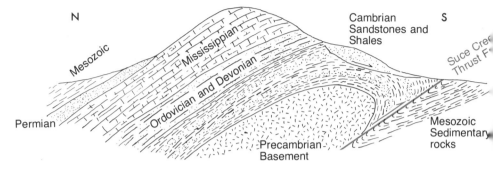

A north-south cross section through the Canyon Mountain Anticline south of Livingston showing how the layered rocks are folded against the Suce Creek Thrust Fault.

meltwater flowing from the end of the glacier. Outwash deposits are flat on top, and look quite unlike the hummocks of the moraines.

About 10 miles south of Livingston, on the east side of the valley, a large terminal moraine blocks the outlet of the Pine Creek valley. This moraine is evidence that a glacier started out of the small side valley, but never flowed very far out into the Yellowstone Valley. Further south, terminal moraines block most of the side valleys coming out of the Absaroka Mountains east of the Yellowstone Valley.

Not far south of Pine Creek, the road steps up over the terminal moraine left by the large glacier that flowed down the valley from the highlands of north-central Yellowstone. This glacier came down the valley during the Pinedale glacial stage, which started about 70,000 years ago, and reached its maximum about 13,000 years ago. Apparently there was an even older glacial period, called the Bull Lake stage, between 160 and 130 thousand years ago. However, the late Pinedale glaciers destroyed almost all of the moraines from the Bull Lake glaciers—at least in this area of the Yellowstone Country. Notice that the entire surface behind the ridge of the terminal moraine is hummocky, evidence that ground moraine till covers the entire valley floor.

Two and one half miles south of Emigrant, a small rock quarry on the west side of the road exposes an excellent example of a basalt flow with glacial striations gouged into its surface during Pinedale glaciation. This quarry is on private land, so seek permission before entering. Basalt is poor in silica and rich in minerals that contain iron and magnesium, which produce the dark color. The post-pile looking columns formed as the liquid cracked and shrank as it solidified. The basalt has been radiometrically dated as about 5.5 million years old.

Dark-colored wall is a dike injected into Eocene volcanic gravel that has been eroded lower than the resistant dike.

At the base of the basalt, you can see some Pliocene stream gravels that in some areas contain horse bones and other mammal remains. Just south of Point-of-Rocks, at Hepburn Mesa, east of the main road, you can see the same basalt flows capping white cliffs of the Pliocene gravels, one of the localities that has produced interesting fossil bones.

Just north of Point-of-Rocks, on the west side of the river, a wall of black rock is a dike injected as molten magma into a crack in horizontally bedded volcanic conglomerates. Because the dike rock resists weathering better than the surrounding volcanic conglomerates, it erodes slowly, and remains as a tall wall.

Not far south of Point-of-Rocks the road enters Yankee Jim Canyon cut by the Yellowstone River into hard Precambrian metamorphic rocks, pink gneiss and dark amphibole minerals. The amphibolites (rocks full of iron-rich amphibole minerals) are so-called because they contain hornblende, a black variety of amphibole. Look for glassy black crystals.

About five thousand years ago a large landslide blocked the entire canyon temporarily damming the river. Some of the large chaotic blocks in the canyon are the remains of this slide. Apparently the dam did not block the river for very long because there is no evidence of lake shorelines or lake silt deposits upstream.

A dirt road leading west into the Tom Miner Basin just north of Yankee Jim Canyon provides access to petrified wood localities of the Sepulcher Formation. Some of the best exposures are at the base of Ramshorn Peak. Unfortunately, collectors have gotten most of the petrified wood at the lower elevations. It is no longer possible to enjoy the experience of seeing the petrified trees without a long hike. If you are willing to climb, you can see some magnificent petrified stumps that remain. One, the "King of Ramshorn" near the top, is a vertical

43

The "King of Ramshorn," a twelve-foot diameter petrified redwood stump near the top of Ramshorn Peak.

redwood over 12 feet in diameter. There is a nice campground near the base for lunch or for a longer visit to the area. If you want to collect some of the wood you must first obtain a permit through the U.S. Forest Service office in Gardiner. While collecting in this area be sure to follow a topographic map and don't stray into Yellowstone National Park. One reason that no collecting is allowed inside Yellowstone National Park is so that future generations as well as our own may enjoy the natural features. Good maps can be purchased in several sport shops in Gardiner.

Devil's Slide on Cinnabar Mountain just south of Yankee Jim Canyon is interesting and photogenic. The rocks are Mesozoic and

The Devil's Slide on the west side of the Yellowstone River south of Corwin Springs. The slide is made of Mesozoic sedimentary rocks tilted up on edge. The soft shales erode faster than resistant sands to make the slide.

Paleozoic sandstones and shales that were originally deposited in horizontal layers, and have since been tilted up on edge. The soft shales erode faster than the hard sandstones and form the low slides. The vertical walls are made of resistant quartz sandstone. The reddish-brown part of the slide is made of the Amsden, Chugwater and Morrison formations.

Above Gardiner, several quarries are mined for travertine to be used for ornamental building stone. Travertine forms around hot springs, as at Mammoth in Yellowstone National Park, where calcium carbonate dissolved from underlying limestone layers by the hot water comes out of solution in the form of hot springs terraces. Often algae help precipitate the limestone and this produces the intricate lacy layering that makes beautiful building material.

Near Gardiner, glaciers carved steep walls in weak volcanic conglomerate that cannot remain as a vertical cliff without the support of the ice. After the ice melted, many of the cliffs slumped into the valley in giant landslides easy to recognize by their chaotic hummocky surfaces, numerous springs and aspen groves, a sharp toe on the valley floor, and a steep scarp where the cliff started to slump. An especially large slide appears southwest of Gardiner upslope from the entrance to Yellowstone National Park.

Hummocky ground from a large landslide at the North Entrance to Yellowstone National Park as seen from Gardiner, Montana.

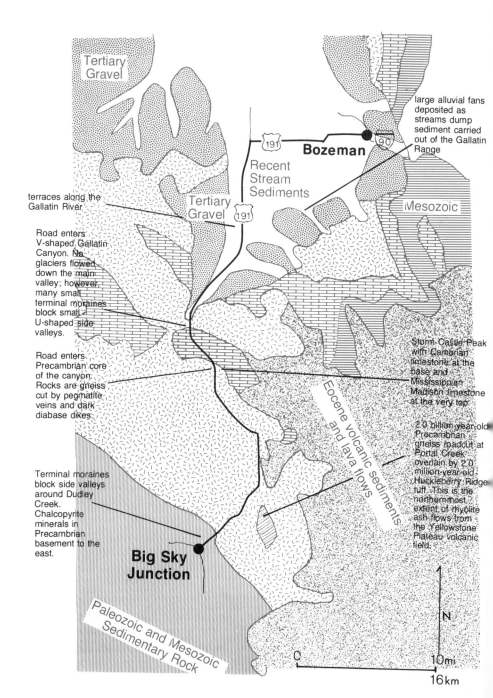

Tertiary Gravel

Bozeman

large alluvial fans deposited as streams dump sediment carried out of the Gallatin Range

Recent Stream Sediments

terraces along the Gallatin River

Tertiary Gravel

Mesozoic

Road enters V-shaped Gallatin Canyon. No glaciers flowed down the main valley; however, many small terminal moraines block small U-shaped side valleys.

Road enters Precambrian core of the canyon. Rocks are gneiss cut by pegmatite veins and dark diabase dikes.

Storm Castle Peak with Cambrian limestone at the base and Mississippian Madison limestone at the very top.

2.0 billion-year-old Precambrian gneiss roadcut at Portal Creek overlain by 2.0 million-year-old Huckleberry Ridge tuff. This is the northernmost extent of rhyolite ash flows from the Yellowstone Plateau volcanic field.

Eocene volcanic sediments and lava flows

Terminal moraines block side valleys around Dudley Creek. Chalcopyrite minerals in Precambrian basement to the east.

Big Sky Junction

Paleozoic and Mesozoic Sedimentary Rock

N

10mi

16km

BOZEMAN — BIG SKY JUNCTION

Bozeman—Big Sky Junction
42 mi./68 km.

Bozeman is at the southern end of a large valley floored with Tertiary sediments, largely volcanic ash deposited in lakes and streams. In places the valley fill is several thousand feet thick. The valleys formed through uplift of the large mountain ranges surrounding the valley. One of these, the Gallatin Range, an uplifted pile of andesitic volcanic material, rises south of town. The road leaves Bozeman on a large alluvial fan made of gravel carried out of the mountains by mudflows and streams. From the sharp turn south on Montana 191 to the Gallatin Canyon, the road crosses stream terraces and glacial outwash of the Gallatin River.

South of Bozeman, the road enters the Gallatin River canyon cut in Precambrian (consult the glossary for definitions of this and other unfamiliar terms) basement rocks. From here to the Big Sky turn-off, all of the outcrops at road level are either recent stream gravels or ancient basement rocks. Most of the Precambrian rocks are banded gneisses with swirly light and dark bands. These rocks formed as older rocks started to flow from very high heat and pressure caused by deep burial. The dark layers contain black amphibole and iron minerals; the pink layers an abundance of quartz, feldspar, and silica-rich minerals. In places, these banded rocks are cut by thin black dikes and white coarse-grained pegmatite veins. The pegmatites formed from the last little bit of liquid in a plutonic rock. Some of the crystals get really huge—from a few inches to several feet across.

If you look up canyon walls, you can occasionally glimpse some of the pale Paleozoic sedimentary rocks that overlie the Precambrian gneiss. One very good exposure is on Stormcastle Peak. The base of the peak is Cambrian sandstone and limestone. The section goes up to the thick cap of Madison Limestone.

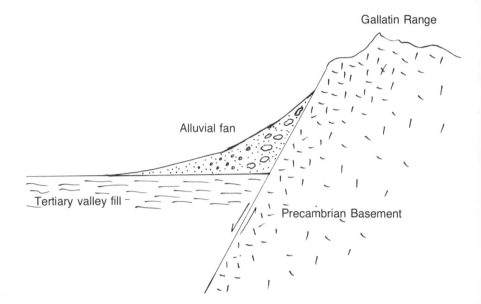

Diagram showing how alluvial fans build out into the valley from the north end of the Gallatin Range near Bozeman.

Notice that the Gallatin Valley and canyon has a distinct narrow profile characteristic of stream-cut valleys, evidence that no glaciers flowed down the main drainage. However, large terminal moraines deposited from glaciers that almost flowed into the Gallatin Valley block the mouths of many small tributary valleys.

At Portal Creek, gneiss in the roadcut has been dated at about two billion years old. This ancient rock is unconformably overlain by an outcrop of 2 million-year-old Huckleberry Ridge Tuff, the northernmost limit of the extensive rhyolite welded-ash flows erupted from the Yellowstone Plateau volcanic field. At Dudley Creek, chalcopyrite (a copper sulfide mineral) and various other ores occur in the Precambrian rocks but not in sufficient quantity to warrant commercial mining. At the same place many terminal moraines block the side valleys. Just north of the entrance to Big Sky is the fault contact between Precambrian basement rock and Paleozoic sedimentary rocks.

Big Sky Junction—West Yellowstone
47 mi./76 km.

From the Big Sky Junction south, the road follows the Gallatin River canyon with Paleozoic sedimentary rocks in the roadcuts. The prominent white ledges are massive Madison Limestone—a marine sedimentary rock that formed in shallow sea water about 300 million years ago. Lone Mountain, near the Big Sky resort, is a glacial horn peak eroded by several glaciers that once flowed in its side valleys.

At Porcupine Creek, a large landslide complex appears west of the highway where it leaves the main part of the canyon. Low rounded hills in the valley are made of black Cretaceous (see glossary or chart on p. vi for ages) marine shales. When wet, clay minerals in the shales absorb water, become very slippery and slide. Use caution during rain storms when driving on dirt roads built on the shales. The dirt road up Taylor Creek provides an interesting short side trip to see a huge landslide that came off the cliffs to the south and blocked the valley. Now Taylor Creek has cut a deep gorge through the toe of the slide. The road crosses the poorly sorted gravel debris of the slide.

The top of Crown Butte on the boundary of Yellowstone National Park about one mile south of Taylor Creek, and on the north side of the road just northwest of Daly Creek, is a remnant of the Huckleberry Ridge Tuff that rests on the non-resistant Cretaceous shales. Notice that you can learn to determine rock type by looking at the slope of a hill. The welded tuff is hard and resists erosion and so holds up a sharp cliff. The Cretaceous shales, on the other hand, are soft and easily eroded into gentle slopes. Lava Butte just to the south is the same type of hill.

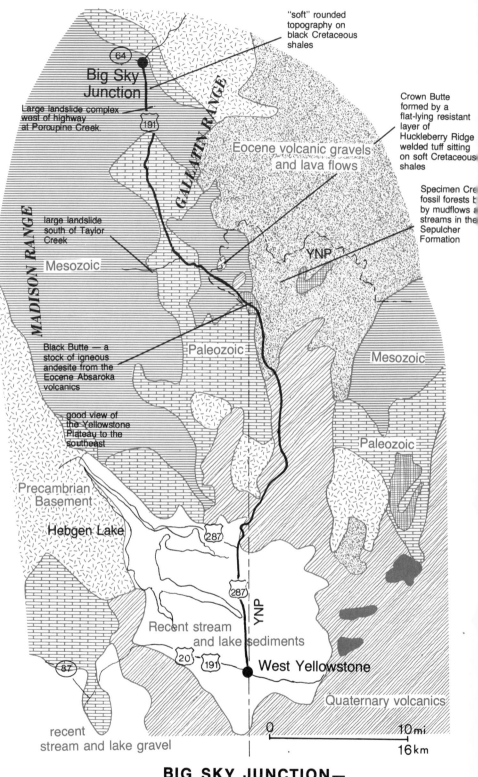

**BIG SKY JUNCTION—
WEST YELLOWSTONE**

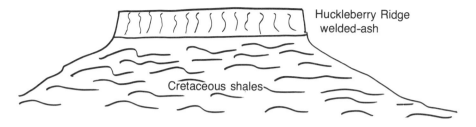

Crown Butte

Huckleberry Ridge
welded-ash

Cretaceous shales

*Cross section of Crown Butte formed from a resistant ledge of 2
million-year-old welded-ash that flowed over Cretaceous shales.*

Black Butte, a volcanic hill between Black Butte Creek and Specimen Creek, is very different from Crown Butte and Lava Butte.
Black Butte formed from magma cooling in the throat of a 50
million-year-old Eocene volcano, rather than from a welded-ash flow.
Rocks that formed in the necks of volcanoes are called stocks, and the
one at Black Butte is made of andesite—a lava intermediate in
chemical composition between basalt and rhyolite. This andesite is
part of the Absaroka volcanic pile that buried the fossil forests described in the Livingston-Gardiner section and in the Introduction.

The Specimen Creek trail head provides access to the Gallatin
Petrified Forest preserved in the Sepulcher Formation, several miles
up the trail, and then up slopes north of the stream; please check with
a park ranger for specific directions. The hike is well worth the
trouble and is less rigorous than that to the fossil forests in the Lamar
River valley. Many upright stumps and prostrate logs are concentrated here; one tall redwood near the top of the first ridge is at least
twelve feet high. Please remember that the exposures along Specimen Creek are within the boundaries of Yellowstone National Park.

*Cross section of Black Butte formed by an Eocene intrusion through
Mesozoic sedimentary rocks.*

Black Butte

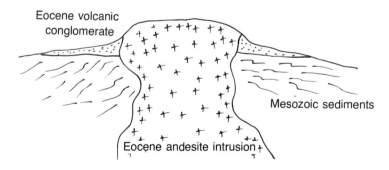

Eocene volcanic
conglomerate

Mesozoic sediments

Eocene andesite intrusion

51

Numerous petrified stumps in the Sepulcher Formation on Specimen Ridge in extreme northwest Yellowstone National Park.

One reason that national parks exist is to protect and preserve natural things—including petrified wood—both for future research and for the enjoyment of all. For these reasons it is illegal to collect specimens from within the park.

South of Specimen Creek, the road is built on Quaternary welded tuffs, Precambrian schists, Paleozoic and Mesozoic sedimentary rocks, and Eocene volcanic conglomerates and andesites. As the road descends a hill, just north of the junction with U.S. 287 to Ennis, a fine panoramic view of the Yellowstone Plateau opens up to the southeast. From this vantage point you can really begin to visualize that the central Yellowstone area is a high flat plateau that represents the filling of an ancient volcanic caldera and the outpouring of welded-ash flows. At this point you are near the edge of older sedimentary rocks uplifted to make the various mountain ranges in the distance. All rocks on the plateau are volcanic lavas and ash deposits erupted during the last two million years. Near West Yellowstone, the road actually enters the Yellowstone Plateau made up of any flat-lying flows of welded-ash flows and lava flows from the three eruptive cycles of the Yellowstone Plateau volcanic field.

West Yellowstone—Norris
28 mi./46 km.

The entire route from West Yellowstone to Norris crosses volcanic rocks erupted during the last 2 million years, and now covered in small patches by stream gravel, lake silt, soil, and hot spring deposits. Some of the most recent flows produced between 150,000 and 80,000 years ago make the valley floor at Swan Lake Flats north of Norris and the canyon walls of the Gibbon Canyon. In fact, all of these volcanic rocks are so young geologically that the body of magma from which they came is still cooling not far beneath the surface. Even though much of the magma may now have solidified, very hot rock still exists only one or two miles down. All stops are to observe the different kinds of lava erupted from the Yellowstone Plateau volcanic field, and to see geysers and hot springs that result when ground water seeps into the earth and comes in contact with these hot rocks.

Apparently, there were at least three cycles of tremendous eruptions of a giant volcanic center that covered all of the Yellowstone region. Radiometric dating of the resulting volcanic rocks shows that explosive volcanism occurred first around 2 million years ago, again at 1.3 million years, and most recently about 600,000 years ago, producing a combined total of almost a thousand cubic miles of rock. Since the last major eruption, there have been several smaller explosions and intermittent lava flows. It is interesting to reflect that times between these major eruptive cycles are about the same as the interval since the last eruption until now. When will the next one be?

Part of the road from Madison Junction to Norris follows the wall of the old caldera. The flat lodgepole pine-covered plateau to the south and east is inside the caldera basin of this volcano. The road crosses the rim near Gibbon Falls about twenty miles east of West Yellow-

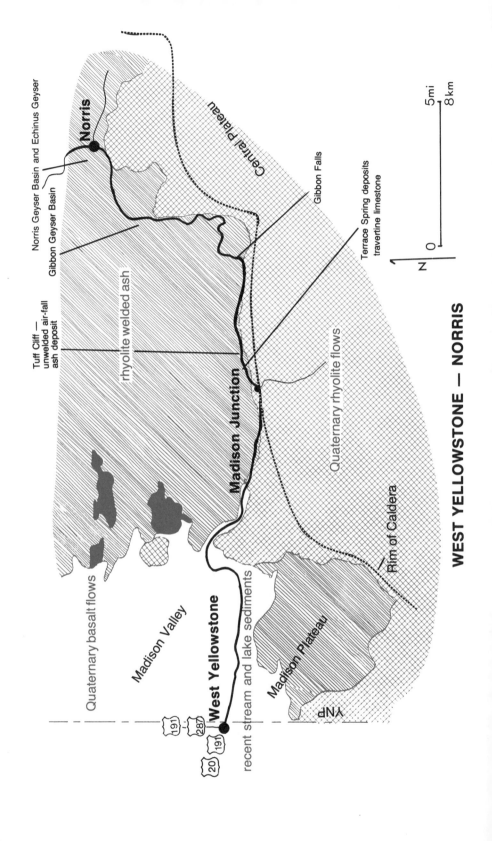

WEST YELLOWSTONE — NORRIS

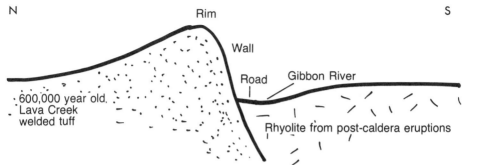

Cross section of the rim of the Yellowstone caldera east of Madison Junction.

stone. The waterfall developed near where the river flows over the very edge of the caldera. Headward erosion caused the falls to migrate upstream from the rim. Good exposures of rock are rare because the dense forest of lodgepole pine obscures most of the bedrock. One exception is the Firehole River canyon just south of Madison (covered in the Madison to West Thumb roadguide) where the river has cut deep between two of the thick rhyolite lava flows that filled the caldera and the Gibbon Canyon south of Gibbon Meadows where the river cuts between rhylite east of the road and welded-ash to the west. From Madison you can also get a good view of a line of hills (Purple Mountain) that rise over 800 feet above the valley to the north and mark the rim of the caldera.

Terrace Springs just east of Madison Junction is an interesting hot springs with a flow of around 1220 gallons per minute. The springs bubble vigorously and appear to be boiling; however, the water is only 144 degrees Fahrenheit and the bubbles are escaping carbon dioxide gas expanding as the pressure drops at the surface. Remember that even though well below the boiling point of water, 144 degrees Fahrenheit is still scalding hot and can cause severe burns. Unlike most of the hot springs in the central part of Yellowstone, this one is depositing both siliceous sinter and travertine limestone. To deposit travertine, the hot water must at some point come in contact with rocks rich in calcium—in most cases a marine limestone. However, most geologists think that because this is at the edge of the old volcano, no limestone rocks could exist—all would have been destroyed by the molten rock. The only rocks beneath this area should be rhyolite lava, a rock rich in silica, not calcium. Thus, this hot springs poses a mystery: Where does the calcium come from? Faults and cracks in the rock may extend out from this area into older limestones and provide a path for hot water saturated with calcium. Other geologists have speculated that the large eruptions did not

Siliceous sinter coats a thermal vent at the Norris Geyser Basin.

destroy all older bedrock. It appears that most lava was produced from ring fractures that circle the caldera complex. As the caldera collapsed, some older bedrock, including limestones, may have been left. The hills south of the Gibbon River are rhyolite flows of the 160,000-year-old Nez Perce flow of the Central Plateau Rhyolite.

Just east of Madison Junction and to the north of the road, Tuff Cliffs exposes volcanic ash. Apparently the ash was not too hot when it came to rest because it is loose and unconsolidated. When very hot ash falls, the individual ash grains weld themselves together to make welded-ash flow, a solid rock.

The rocks exposed on the east wall of the Gibbon Canyon are plateau rhyolites of the Gibbon River flow erupted from a vent along the caldera rim about 87,000 years ago. North and west of the road are welded-ash flows of the Lava Creek Tuff (erupted from the Yellowstone caldera 600,000 years ago) that lie outside the caldera rim.

There are quite a few hot springs between Madison and Norris. Beryl Springs, eight miles from Madison Junction, shows extensive siliceous sinter deposits. The greatest concentration of thermal features is in Gibbon Meadows about nine miles north of Madison. This area, called the Gibbon Geyser Basin, has a few geysers and some interesting hot springs. Artist Paint Pots, reached by a mile round-trip hike from a trail at the southern end of Gibbon Meadows, has bright colors from a concentration of iron oxide minerals. Mud pots in the area show that this spot must have a limited water supply so the clays don't get washed away. Chocolate Pots, along the highway, get their color from a very high concentration of around 50% iron oxide in addition to some aluminum oxide and manganese oxide. The temperaure at Chocolate Pots is only around 130 degrees Fahrenheit, cool compared to other basins in the park.

Norris Geyser Basin is one of the best places to see various kinds of

Bubbles bursting in mud pots near Artist Paint Pots make interesting fried-egg shaped patterns. Mud pots form where there is too little groundwater to wash away rock altered by the hot acidic water.

geysers, hot springs and fumaroles. Unfortunately, this area is often overlooked in favor of Old Faithful and the Upper Geyser Basin. Many of the geysers and hot springs in the Norris Geyser Basin deposit terraces of silica called siliceous sinter. Hot water dissolves silica from underlying lava and ash flows; then as the water cools near the surface, the silica comes out of solution to form the sinter terraces. One reason that a lot of silica is deposited at Norris is because the basin has the hottest ground water of any thermal area in the park. One drill hole found 465-degree-Fahrenheit water only 1,100 feet below the surface, and hotter water surely exists below drill depth. Surface water can reach 199 degrees Fahrenheit (the temperature that water boils at this elevation), or hotter, around some of the geyser pools. This high temperature allows the water to contain quite a bit of dissolved silica, which precipitates as sinter.   57

Terraces of siliceous sinter at Cistern Springs in the Norris Geyser Basin.

Some of the geysers and fumaroles produce superheated water (water above the boiling point) during part of their cycle. One fumarole in the Norris basin has vented steam at a temperature of around 270 degrees Fahrenheit.

Be careful when walking through the basins to keep spray from the geysers off your glasses and camera lens. When the water evaporates it will leave permanent spots of siliceous sinter on your lens, which are impossible to remove without damage to the glass. Also heed the park signs and stay on the boardwalks so that you don't break through an unstable sinter crust and receive a severe burn. Every year people ruin their vacations by ignoring this logical rule. Also,

Cross section of a typical geyser. Rain water soaks deep into the ground and is heated by hot volcanic rock. Weight of the water column and possibly a winding joint system to the surface causes high pressures to develop before the steam shoots out in a geyser.

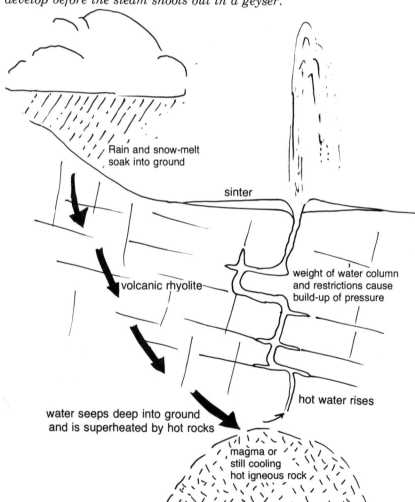

Rain and snow-melt soak into ground

sinter

weight of water column and restrictions cause build-up of pressure

volcanic rhyolite

hot water rises

water seeps deep into ground and is superheated by hot rocks

magma or still cooling hot igneous rock

several once-beautiful geysers in the basin have been destroyed by rocks and trash thoughtlessly thrown into the pools; please don't do this or remove specimens.

Besides being the hottest, the Norris Geyser Basin is also the most shallow of any in the park, with very hot rock near the surface. High pressures allow liquid water to reach at least 527 degrees Fahrenheit at depth before it boils. When this superheated water finally boils, the resulting 1500-time expansion blasts some of the water out as geysers.

I think Echinus Geyser in the Norris Geyser Basin is one of the best in the park to observe how geysers work. Ground water from the surrounding area seeps deep into the earth where the hot volcanic rock heats it. This hot water rises and fills a long vertical tube, usually a joint or crack in the rock, that reaches the surface. The water in contact with the hot rock becomes so hot that it would boil if it were not for the pressure of the weight of the water already in the tube. Eventually the water becomes so superheated that it expands and pushes some of the water out. This reduces the pressure at depth, and the entire column flashes into steam and pushes the water and steam out into a geyser. Water then runs back into the tube and the process starts again. Some geysers erupt like clockwork. Echinus Geyser erupts every 30-90 minutes and is well worth the wait. From a vantage point directly above the pool you can watch the entire sequence. Just before Echinus erupts, the pool slowly fills with water from the expansion. Bubbling then relieves the pressure, the column boils, and the eruption begins. After the main eruption, stay and watch the water drain back into the tube at the base of the pool to start the sequence over again. For additional information on the numerous geysers and pools in the Norris Geyser Basin, visit the small museum where exhibits and a small brochure are available. You may also wish to read the excellent descriptions of each geyser in *The Geysers of Yellowstone* by Bryan (1979), sold in the park.   59

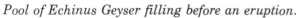

Pool of Echinus Geyser filling before an eruption.

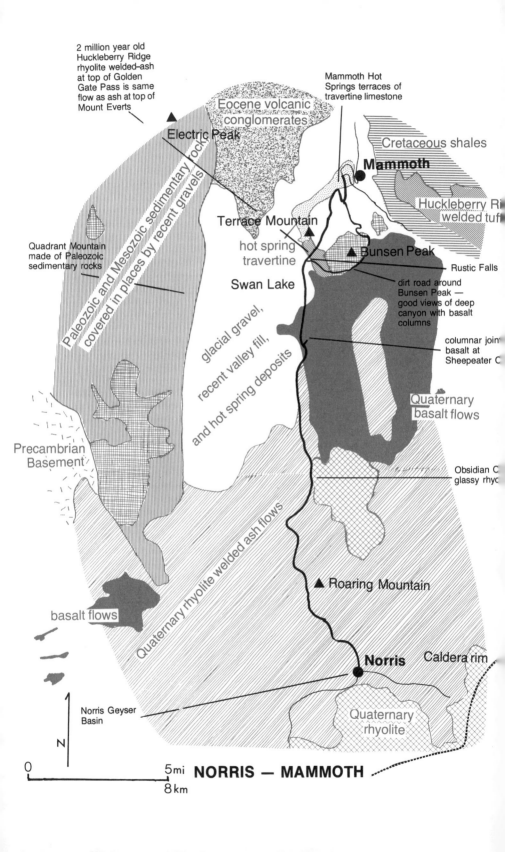

2 million year old Huckleberry Ridge rhyolite welded-ash at top of Golden Gate Pass is same flow as ash at top of Mount Everts

Mammoth Hot Springs terraces of travertine limestone

Eocene volcanic conglomerates

Electric Peak

Cretaceous shales

Mammoth

Huckleberry Ri welded tuf

Terrace Mountain

hot spring travertine

Bunsen Peak

Rustic Falls

Quadrant Mountain made of Paleozoic sedimentary rocks

Paleozoic and Mesozoic sedimentary rocks covered in places by recent gravels

Swan Lake

dirt road around Bunsen Peak — good views of deep canyon with basalt columns

columnar joint basalt at Sheepeater C

glacial gravel, recent valley fill, and hot spring deposits

Quaternary basalt flows

Precambrian Basement

Obsidian C glassy rhyc

Quaternary rhyolite welded ash flows

basalt flows

Roaring Mountain

Norris

Caldera rim

N

Norris Geyser Basin

Quaternary rhyolite

0   5mi   **NORRIS — MAMMOTH**
                     8 km

Norris-Mammoth
21 mi./34 km.

Roaring Mountain, about five miles north of Norris, is the remains of a steam explosion that happened sometime during Pinedale glaciation. The mountain used to "roar" (it now only hisses and rumbles) because of escaping gases and steam from a boiling ground water reservoir at shallow depth. The thermal activity here, as in all areas of the park, changes through time. The roar of the mountain has declined greatly since it was first described by explorers in the 1800s.

Even though not visible from the road, there is another fascinating hydrothermal explosion crater on a flat mesa or bench above Roaring Mountain. The caprock is made of one of the welded tuff units. Some scientists have proposed that a very large crater in the middle of the bench formed from a steam explosion when confining pressures were released quickly, possibly from fractures caused by an earthquake. This pressure release allowed flash boiling of very hot water (possibly 480 degrees Fahrenheit or hotter) in the entire water table. Angular blocks and mud surround the crater, and some debris was thrown over half a mile from the resulting expansion of water flashing into steam.

Obsidian Cliff, eight miles north of Norris, is well worth a short stop. The cliffs are made of a lava flow extruded around 180,000 years ago, during the post-caldera phase of the third eruptive cycle. The large columns at the base of the cliff across the road from the parking area formed as the lava cooled and shrank. The rubbly zone near the top of the flow is a breccia that formed when the surface hardened, but the liquid core continued to move. Large folds near the top of the cliff are pressure ridges from buckling on the surface of the flow.

Flow bands in a chunk of obsidian of the Roaring Mountain member of the Plateau Rhyolite at Obsidian Cliffs. The nickel is there for scale.

Obsidian is a glassy rock with no crystals that forms when dry rhyolite magma hardens quickly. Geologists have speculated as to why obsidian has no visible crystals; none can be seen even under a very high powered microscope. It was once assumed that the liquid rock must have hardened very quickly—possibly because the lava flowed against the side of a glacier or a valley wall. In general, rocks that cool quickly do have smaller-sized crystals than those that cool slowly because it takes a lot of time for a large crystal to grow. If the rock instantly solidifies, all molecules are "frozen" in place before they can arrange themselves into crystals. This does not completely explain Obsidian Cliff. For one thing, even though the flow has a crystalline core, the fine-grained margin is very thick; thus, the thick rind would have to have cooled at the same rate. This is hard to imagine because rock is an extremely good insulator and even a glacier might not cool the rind fast enough.

An alternate explanation to the chilled margin has been suggested by some geologists. Chemical analyses of the obsidian show that the lava was very dry with little dissolved water. Without water, no crystals could form, leaving a glassy rock regardless of how long it took to cool. If much water had been dissolved in the magma, it should have blasted into steam making a welded-ash rather than a flow.

Because obsidian is very dry, it can be used in an interesting way to date early human artifacts. After it cools, the lava starts to absorb moisture from the atmosphere at a fairly constant rate. When humans chipped out a spear or an arrow point, they flaked off the thin skin of rock that had absorbed water. Then a new rind begins to form. By measuring the thickness of the rind produced by water on a chipped surface, scientists can then tell how long ago a point was fashioned.

Lithophysae in obsidian at Obsidian Cliffs.

Several blocks of the obsidian along the road show bands that formed as the magma flowed. Concentric ovoid spaces in the obsidian in the blocks at the base of the cliff are "lithophysae," gas cavities in the liquid magma that later became lined with small crystals. Cavities completely filled with radiating crystals are called spherulites. As you scramble around the base of the cliff to see the things I have just described, remember not to collect samples—that is against the law and you would be subject to arrest.

Two kinds of lava erupted from the Yellowstone Plateau volcanic field and both appear a few miles south of Mammoth near the Golden Gate. Basalt is a black fine-grained volcanic rock that forms from the cooling of lava that is low in silica and fluid enough to flow easily over long distances. In contrast, rhyolite typically forms a light pale rock rich in silica. However, the extensive layers of rhyolite outside the central caldera area of Yellowstone formed in a slightly different fashion than the layers of dark basalt.

When viscous rhyolite erupts it piles up like bread dough. Thus, it

63

Cross section of a six-sided column of basalt. Lens cap indicates scale.

does not form such extensive thin layers as exist around the margins of the Yellowstone Plateau. Apparently when the volcano exploded, the gases and explosive force shattered the magma into millions of tiny particles called volcanic ash. These ash grains mixed with superhot volcanic gas and flowed as a dense cloud over the landscape. The ash cloud was so hot that when it came to rest, the ash grains welded themselves into solid rock.

Sheepeater Cliffs, down a short side road to the east less than a mile north of Indian Creek Campground, is an excellent place to observe the dark basalt flows with their long vertical columns of the cliff formed from shrinkage as the magma crystallized and cracked. Along the Gardiner River, just downstream, is a complex area of basalt and rhyolite; two types of liquids appear to have mixed. You can see the mixed zone along the riverside where there is rock intermediate in color and composition between basalt and rhyolite.

North of the side road to Sheepeater Cliffs, a pull-out on the west side of the road at Swan Lake Flats provides an exhibit explaining the geology of uplifted Paleozoic and Mesozoic rocks to the west. They make up the southern end of the Gallatin Range that you can see before you to the west. Glacial ice was at least 2200 feet thick here during the Pinedale glaciation and covered all of the peaks in the Washburn Range behind you to the east. Rock surfaces on the tops of some of these mountains have glacial striations, showing that they were once covered by the ice. Hot springs existed in the region long before the glaciers came. Terrace Mountain at the north end of Swan Lake Flats, is an old hot spring deposit covered by glacial gravel. Large streamlined piles of glacial till, called drumlins, cover part of the flats.

Immediately before the main road descends into Golden Gate, an unpaved one-way 4-mile-long road to the east allows a view of some interesting rocks and scenery. This is a rough road with tight curves and is not recommended for motor homes or trailers. The road loops around the east side of Bunsen Peak and returns to the main road just

Basalt columns i
flow of Swan Lak
Flat Basalt at
Sheepeater Cliffs.

Two million-year-old Huckleberry Ridge welded-ash at Golden Gate. This is the same flow that exists at the top of Mount Everts visible in the central background of the photo.

south of Mammoth. Along the way are views of a spectacular canyon along Sheepeater Cliffs made of long, thin basalt columns and a trail head for Osprey Falls.

Light-colored rhyolitic welded-ash flows are exposed in the vicinity of Golden Gate where the road starts downhill. A parking area on the right, in the pass, is a good place to look at the flows exposed on the other side of the road. Notice the small waterfall (Rustic Falls), on the same side of the road as the parking area, that formed because the rhyolite resists erosion and produces a vertical cliff. This welded-ash (or "welded tuff") is the Huckleberry Ridge Tuff that formed during the first eruptive cycle of the Yellowstone Plateau volcanic field 2 million years ago. Look northeast to see that this is at the same level as the welded-ash at the top of Mount Everts (see the Gardiner to Tower Junction roadguide) and is, in fact, part of the same welded-ash flow that once was continuous across what is now a valley. The welded tuff was dissected by streams and other processes that carved the modern valley in the last 2 million years, after the Huckleberry Ridge Tuff cooled.

Along the descent into the Mammoth area, you pass through an area of large jumbled blocks, called "the Hoodoos" or Silver Gate, made of ancient hot spring deposits mixed with blocks of welded tuff.

These blocks tumbled down from Terrace Mountain in a large land-slide that slid on weak underlying Cretaceous shale. Please note that even though the proper name for this area is "the Hoodoos," these are not proper hoodoos that form when a resistant cap rock keeps a soft shaft from eroding away.

A side road and several trails lead onto the Upper Mammoth Hot Springs Terraces. Sedimentary limestone layers are exposed at the surface. Surface water soaking into the ground becomes heated and dissolves calcium carbonate from the rock. Carbon dioxide gas escapes from solution as the pressure drops near the surface, and calcium carbonate deposits on the hot springs pools in the form of travertine limestone. The water here appears to have a free path to the surface because it flows continuously and freely from the springs, and is not hot enough to make a geyser.

Apparently the hot ground water flows here from the Norris Geyser Basin about 20 miles to the south, allowing it to cool. This area is also much farther away from the remnants of the magma chamber than the geyser basins. Drilling in the Mammoth area revealed that the ground water, even at depth, is only around 167 degrees Fahrenheit—not hot enough to boil water. All of the bubbling in the hot springs is from carbon dioxide gas expanding at the surface, not boiling water. The carbon dioxide is dissolved in the water from the underlying limestone and comes out of solution as the pressure lessens at the surface, just as when you shake a can of carbonated soda and then pop the top. For more information on the over-100 hot springs in the Mammoth area, consult the pamphlet by Bargar (1978), available in the park research library at Mammoth, but out of print.

Travertine terraces at Mammoth Hot Springs.

Gardiner—Tower Junction
23 mi./37 km.

The road from Gardiner to Tower Junction passes exposures of a variety of geologic features. These can be broken down into six types. 1) Marine shales of Cretaceous (see Glossary for this and other unfamiliar terms) age exposed in the vicinity of Mammoth on Mount Everts; 2) hot spring deposits around Mammoth formed by the dissolution and reprecipitation of calcite from older underlying formations; 3) rhyolitic ash and basaltic lava flows 600,000 to 2 million years old from eruptions on the Yellowstone Plateau volcanic field; 4) Eocene conglomerate that preserves various formations of the fossil forests; 5) Pinedale glacial till; 6) Precambrian basement rocks from the Beartooth uplift.

If you look almost due west from the Gardiner area, you see spectacular views of Electric Peak, a glaciated horn that is the highest summit in the Gallatin Range. Electric Peak (elev. 10,992 feet), and the lower Sepulcher Mountain (elev. 9,652 feet), are made of Cretaceous sedimentary rocks that contain Eocene igneous intrusions. These two mountains must have formed one of the volcanic vents of the Absaroka Volcanic Supergroup that buried the fossil forests.

From Gardiner, the road into the park crosses the toe of a huge landslide that came from the steep cliffs exposed on the high mountain to the southwest. Notice the irregular hummocky surface of this flow, easily visible while looking south from downtown Gardiner or the park entrance. Along the edge of the Yellowstone River, the toe of this slide has been eroded back into the river terraces.

Quarries for ornamental stone on the ridge above Gardiner are mined for travertine. It formed in a now extinct hot spring system that must have been similar to the one still active at Mammoth Hot Springs.

67

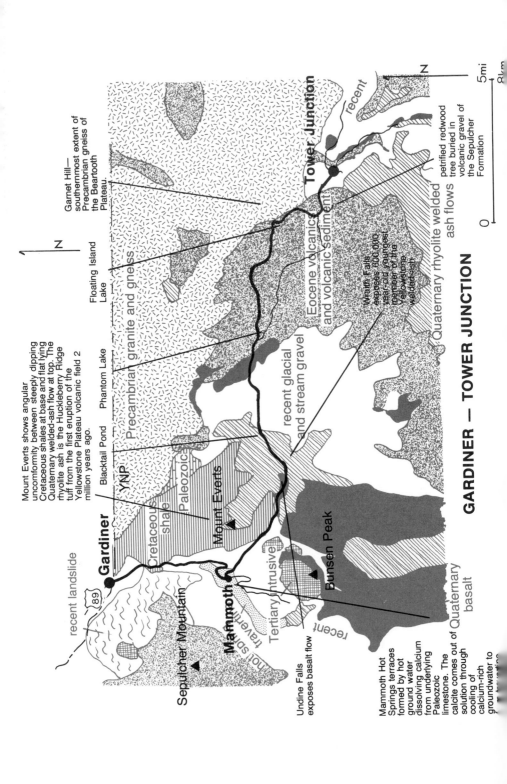

Mount Everts shows angular uncomformity between steeply dipping Cretaceous shales at base and flat lying Quaternary welded-ash flow at top. The rhyolite ash is the Huckleberry Ridge tuff from the first eruption of the Yellowstone Plateau volcanic field 2 million years ago.

Garnet Hill— southernmost extent of Precambrian gneiss of the Beartooth Plateau

Floating Island Lake

Phantom Lake

Blacktail Pond

YNP

Precambrian granite and gneiss

Paleozoic

Cretaceous shale

Mount Everts

Gardiner

recent landslide

89

Sepulcher Mountain

Mammoth

hot springs travertine

Tertiary intrusive

recent

Bunsen Peak

Undine Falls exposes basalt flow

recent glacial and stream gravel

Eocene volcanic and volcanic sediment

Tower Junction

recent

Wraith Falls exposes 600,000 year-old youngest member of the Yellowstone welded-ash

Quaternary rhyolite welded ash flows

petrified redwood tree buried in volcanic gravel of the Sepulcher Formation

Quaternary basalt

Mammoth Hot Springs terraces formed by hot ground water dissolving calcium from underlying Paleozoic limestone. The calcite comes out of solution through cooling of calcium-rich groundwater to

N

N

0   5mi

0   8km

GARDINER — TOWER JUNCTION

Gouged out glacial trough on Electric Peak (elev. 10,992')–the highest peak in the Gallatin Range.

South of Gardiner, along the Gardiner River, the road enters a narrow canyon where a large landslide blocks the valley. High cliff faces east of the road are black shales, sandstone ledges and gravel lenses of, in order from bottom to top, Cody Shale, Telegraph Creek Formation, Eagle Sandstone, and Everts Formation (see the stratigraphic list of formations on p. vi). These formations together total over 3500 feet thick and were deposited in and along the edge of an ocean some 100 million years ago during Cretaceous time. Rocks near the base of the cliff are marine formations deposited in water on a shallow continental shelf and on beaches. Near the top of the cliffs the Everts Formation consists of sediments accumulated on a delta. There are lenses of stream sand and gravel, coal, and black lagoon shales in this formation. A modern analogy for these Cretaceous rocks is the Mississippi River delta. There, rivers carry continental sediments and pile them up in the form of a delta on shallow marine rocks of the Gulf of Mexico.

Where the road starts to climb out of the canyon at the sign marking the 45th parallel, a trail leads upstream along the Gardiner River to a hot spring pool called Boiling River. This spring has one of the highest flow rates of any thermal feature in the park and is one area where you can soak in the spring-heated river water. Be careful not to get into the spring itself or you might get badly burned—or receive a fine because it is illegal to bathe in any hot spring in the park. Notice the travertine limestone deposits around the edges of the pools. Also note that this area is closed during the posted hours that change seasonally.

The small conical knob (informally called "Dude Hill") about a mile south of the 45th parallel sign and across the road from the campground, is a deposit of glacial gravel. The hill may have formed

69

when gravel filled a hole melted into glacial ice by a hot spring about 15,000 years ago.

Between the Mammoth campground and Mammoth proper, is a beautiful example of an ancient hot spring deposit in a roadcut. The layers precipitated out of solution in a hot spring similar to Mammoth Hot Springs just up the road. Like their modern counterparts, rocks deposited around ancient hot springs are called travertine or algal tufa. It was once assumed that all travertine formed inorganically, as calcium carbonate dissolved in the hot water was precipitated out of solution as the water cooled. However, some geologists now think that mats of bacteria and algae living in the hot water aid deposition of travertine in some of the springs.

Mount Everts east of Mammoth showing an angular unconformity. Layers tilting down to the left are Cretaceous shales and make up most of the cliffs. Near the top, these are overlain by a thin horizontal resistant ledge of 2 million-year-old Huckleberry Ridge welded-ash. The shales were tilted and eroded before the ash was deposited. The modern valley was cut after the ash was emplaced.

Where the road levels out at the top of the hill there is a pull-off on the right that provides a safe place to stop and view some interesting rocks to the east on Mount Everts. These same rocks can also be seen from the hot springs and from anywhere along the first several miles of road leading toward Tower Junction, or from Golden Gate to the south, up the hill toward Norris. Notice that the Cretaceous shales making up the bulk of Mount Everts dip at 10-20 degrees to the west. Because most sedimentary rock was deposited in more-or-less hori-

Minerva Terraces.
National Park Service Photograph.

zontal layers, we can surmise that these have been tilted by mountain building forces after they were deposited. Notice, however, that at the very top of the mountain there is a dark brown rock layer that is horizontal and overlies the older rock in an angular unconformity. This younger rock is the Huckleberry Ridge Tuff, a welded-ash flow erupted from the Yellowstone Plateau volcanic field about 2 million years ago, the oldest major eruptive unit of the region. Obviously, the tuff erupted after the older rocks beneath them were hardened, tilted, and then eroded into a flat surface.

The Mammoth Hot Springs provide a good place to observe numerous geologic features. From even a casual inspection of the Mammoth area it is apparent that there are many hot springs but no geysers. Both of these thermal features form as snow-melt and rain water seep deep into the crust of the earth and become heated by very hot volcanic rocks.

Evidently, the Mammoth area has open cracks, because the hot springs and pools flow continuously. There are no geysers at Mammoth, because the water is not hot enough. The water is cooler than many other areas in Yellowstone, even at depth, because it has flowed to Mammoth from the Norris Geyser Basin, about 20 miles to the south. As the hot water rises to the surface, it comes in contact with many layers of rock. Some of these layers, such as limestone, dissolve readily in hot water and the water becomes saturated in calcium carbonate.

As this mineral-rich water cools and the pressure drops near the surface, some of the minerals precipitate out of solution as calcium carbonate deposits on the hot spring terraces. Most of the mineral is deposited by evaporation of some of the water at the surface and loss of carbon dioxide gas. Many species of algae and bacteria live in the warm waters of the hot springs and their life processes can also cause limestone to precipitate out of solution. When some of the hot spring terrace rocks are sliced thin and observed under a microscope, the rock reveals many very thin lacy layers recording days when photo-synthetic bacteria grew, and nights when they rested. Thus, the fascinating forms around the Mammoth pools are a result of hot water dissolving and reprecipitating calcite from older rock, and of bacteria and algae.

Deposition of travertine occurs rapidly. Geologists have reported an average rate of about eight inches of rock deposited each year. A drill hole by the U.S. Geological Survey found 253.4 feet of travertine in the Mammoth area. The maximum temperature encountered by this drilling in the hot springs is only around 167 degrees Fahrenheit—quite cool (still hot enough to cause the most severe burns!) as compared to the subsurface water temperatures of the other thermal basins in the central part of the caldera.

From Mammoth east, the road crosses Pinedale glacial till before crossing a deep canyon of the Gardiner River. Moraines near the Gardiner River bridge were deposited near where two Pinedale glaciers joined, one flowing from the Gallatin Range and the other flowing west from the upper Lamar River valley. Downstream from the junction, and in the moraines near the bridge, are erratic boulders carried from the Gallatin Range. On Bunsen Peak south of the canyon there is a distinct horizontal line dividing older from younger trees that have grown since this area was burned in a large forest fire in 1886, but has since become reforested. About a mile and a half from the Gardiner River bridge, a sharp-crested lateral moraine follows the downhill side of the road. The moraine formed along the edge of the glacier that flowed from the Beartooth uplift and the Upper

72

Travertine terraces at Minerva Springs, Mammoth Hot Springs.

Lamar River. Bunsen Peak to the southwest was covered by at least 750 feet of ice during the Pinedale glaciation.

Undine Falls, one of a series of two or three waterfalls along a short stretch of Lava Creek, 2.5 miles east of the bridge and 4.5 miles from Mammoth and to the north (left) of the road, flows over hard layers of a 700,000-year-old Pleistocene basalt lava flow. The flow must have come down an ancient stream channel because it now fills a valley that was eroded into the soft underlying Cretaceous shales. The falls exist because the stream can't erode basalt as fast as soft shale, so the basalt forms a steep cliff, called a nick point.

For the next 15-20 miles, the road follows various rhyolite welded-ash and basalt lava flows erupted from the Yellowstone Plateau volcanic field. At Wraith Falls, 5.5 miles east of Mammoth and to the south of the road, the creek exposes the youngest of the major rhyolitic units (about 600,000 years old), called the Lava Creek Tuff.

Blacktail Pond, a little over six miles east of Mammoth, fills an ancient glacial channel carved by meltwater. Cores of sediment from the pond contained thin bands of volcanic ash that drifted here from the eruptions of Mount Mazama (Crater Lake, Oregon) 6,600 years ago and of Glacier Peak in the North Cascade Mountains 13,000 years ago. The remains of plant pollen grains in the lake sediment show a succession of plant types as the climate changed over the last 11,000 years. However, the climate must not have changed very much because all of the fossil species are of plants that still grow today in Yellowstone Park. Botanists thus argue that there have been no major climate changes since eruption of Glacier Peak. At Phantom Lake, just under ten miles east from Mammoth, glacial meltwater cut deep channels into Eocene basalt flows, the 48 million-year-old Crescent Hill Basalt. Some of these ice-carved channels can also be seen along the Blacktail Plateau drive.

Twelve to thirteen miles east of Mammoth, and six miles west of Tower Junction, there are some pull-outs along the north side of the road allowing excellent views up the valley to the northeast. Low rounded pinkish to white hills, such as Garnet Hill and Hellroaring Mountain, are Precambrian granite and gneiss, the southernmost exposures of basement rock from the Beartooth uplift (see Cooke City—Roosevelt roadguide). Brown cliffs in the farther distance are Eocene volcanic conglomerates of the Absaroka Volcanic Supergroup, complete with fossil forests. One very good area to see the view I have just described is a pull-out where you can almost see the junction of Hellroaring Creek with the Yellowstone River north of the road. This junction is at the southern edge of the Precambrian Beartooth block.

Near Floating Island Lake, about 16 miles east of Mammoth, there is an unpaved road with a trail head sign on the north (left side of the road) that leads one quarter mile to an old gravel pit built at the base of a scree slope. The scree blocks are falling from a cliff of welded-ash from one of the Absaroka volcanoes. The welded-ash is about 48-49 million years old and is called the Lost Creek Tuff. It differs from most of the Absaroka volcanic deposits because it contains abundant silica—almost as much as the Quaternary rhyolites, but not quite. It is named either dacite, or trachyte.

One and a half miles west of Roosevelt, about 18.5 miles from Mammoth, a sign points to the Petrified Tree road. This is the only place in Yellowstone where you can actually drive to one of the petrified trees. Heed the sign and don't take trailers or large motor homes up this side road unless you want to leave them there, since it would be extremely difficult to turn around. The large standing petrified tree is a redwood anatomically indistinguishable from modern redwoods growing today along the California coast. Photographs from the late nineteenth century show an additional tree standing next to this one, and written reports tell of many other horizontal logs as well as upright stumps. Collectors have since gotten just about every scrap. All that remains are depressions in the soil. This tree is buried in water-reworked volcanic ash of the Sepulcher Formation, the same one that buries petrified trees in Tom Miner Basin and Specimen Creek in the extreme northwest corner of the park.

Immediately behind the buildings of Roosevelt Lodge, there are some glacial and stream gravels. One ridge is what remains of a pile of rocks and debris that rode on the edge of an ancient glacier. As the glacier melted, it dropped this long ridge of rocks.

Tower Junction—Cooke City
32 mi./52 km.

The road at Tower Junction is on glacial sand and gravel. Just where the road descends into the river canyon, a basalt flow underlies some of the gravel. The small flat-topped hill to the north and right across the road from the picnic area is Junction Butte, a remnant of the 2 million-year-old Junction Butte basalt. The same basalt flow is exposed along the side of the road between Tower Junction and Tower Falls. It filled an ancient stream channel. Where the road crosses the north end of the Grand Canyon of the Yellowstone River, you may wish to pause and look upstream into the canyon. Rocks near the river level, seen from the bridge, are Eocene conglomerates altered by hot spring water to various shades of yellow and orange. Usually several small steam vents are visible from the bridge, and you can detect a strong stench of sulphur if the winds are blowing downstream.

Farther northeast, the road follows the lower Lamar River valley. From here to Cooke City the valley of the Lamar River and its various tributaries course through the most scenic and least visited (at least along a road) part of the park. Wildlife is almost always abundant during fall through spring because many animals winter here in this valley to escape the harsh climate in the high meadows.

Rocks along the entire route are mostly the Lamar River Formation, and contain abundant petrified trees. All of the numerous dark brown cliffs contain excellent fossil forests. Minor additional rock types are 2.7 billion-year-old granite and gneiss in the Lamar Canyon just north of Slough Creek, and some 300 million-year-old Paleozoic limestone at the sharp bend where Soda Butte Creek joins the Lamar River. Both of these areas are excellent places to stop and look at the

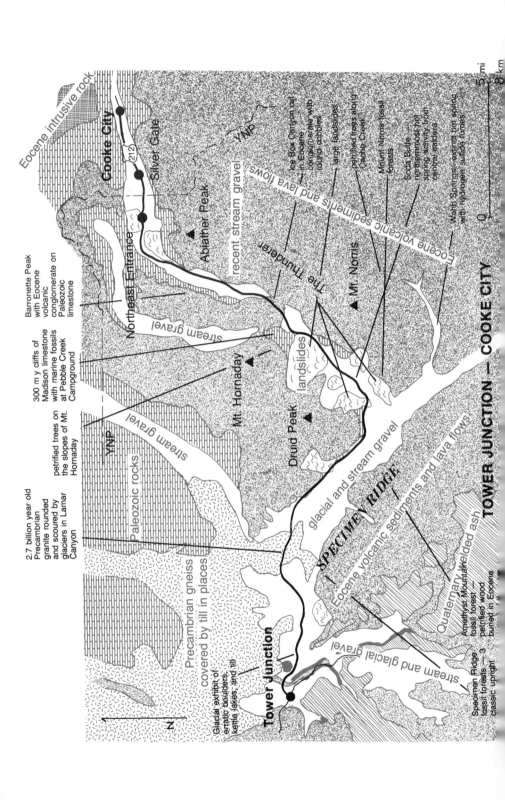

TOWER JUNCTION — COOKE CITY

2.7 billion year old Precambrian granite rounded and scoured by glaciers in Lamar Canyon

petrified trees on the slopes of Mt. Hornaday

300 m y cliffs of Madison limestone with marine fossils at Pebble Creek Campground

Barronette Peak with Eocene volcanic conglomerate on Paleozoic limestone

Eocene intrusive rock

Cooke City

212

Silver Gate

Northeast Entrance

YNP

Abiathar Peak

recent stream gravel

Eocene volcanic sediments and lava flows

Ice Box Canyon cut in Eocene conglomerate with rounded cobbles

large landslides

petrified trees along Cache Creek

Mount Norris fossil forests

Soda Butte — northernmost hot spring activity from central caldera

Wahb Springs extinct hot spring with hydrogen sulfide fumes

The Thunderer

Mt. Norris

stream gravel

Mt. Hornaday

landslides

Druid Peak

glacial and stream gravel

SPECIMEN RIDGE

Eocene volcanic sediments and lava flows

Paleozoic rocks

stream gravel

YNP

Precambrian gneiss covered by till in places

Glacial exhibit of erratic boulders, kettle lakes, and till

Tower Junction

N

stream and glacial gravel

Quaternary welded ash

Amethyst Mountain fossil forest — petrified wood buried in Eocene

Specimen Ridge fossil forests — 3 classic upright

0   5 mi
0   8 km

ancient rocks on which the Eocene sediments are deposited. Because the 50 million-year-old volcanic gravel rests directly on rocks ranging from Precambrian to early Paleozoic, .3-2.1 billion years of earth's history is missing in a giant unconformity. The rocks in the canyon east of Slough Creek are light pink banded gneiss, a rock formed by extreme heat and pressure causing complete recrystallization of even older rocks. The rocks at the Paleozoic outcrop are layered sedimentary rocks formed in a shallow ocean; they contain many fine examples of invertebrate fossils.

As the road breaks out of the canyon north of the Yellowstone River into the valley, many exceptional examples of the work of ancient glaciers appear. A pull-out on the north side of the road at the base of Junction Butte along with a park interpretive display explain many of them. Ken Pierce of the U.S. Geological Survey has shown that Pinedale glaciers flowed down this valley from the north about 13,000 years ago. The glacier must have been over 90 miles long from the head on Granite Peak, outside the park area on the Beartooth Plateau, to the toe in the Yellowstone Valley south of Livingston. All of Specimen Ridge and most of the nearby mountains were covered by ice that reached an elevation of 9,000 feet.

Large erratic granite boulders strewn across the valley were carried by the ice from Precambrian outcrops in the Lamar Canyon north of Slough Creek. Notice how small Douglas fir trees grow on the north sides of the boulders to take advantage of shade and resulting moist soil to survive in the dry valley. Small swampy kettle ponds formed where large blocks of glacial ice broke off the main glacier, and were covered by meltwater-deposited sand as the glacier receded. When the ice finally melted, the ground collapsed and filled with water forming the kettle lakes. Additional evidence of the glaciers appears in the rounded and striated or grooved cliffs in the canyon. These were scoured by pebbles trapped in the glacial ice that acted just like sandpaper to erode the granite and gneiss.

Numerous mudflows and landslides along the Lamar River and

Glacial erratic granite boulder at the base of Junction Butte. Trees grow on the shaded sides of the large boulders.

A kettle lake with huge granite erratic boulders near Slough Creek. The second rounded ridge behind the lake is made of Precambrian granite. The dark cliffs forming the skyline in the background of the upper left are Eocene volcanic gravel.
Photo by Dave Alt.

tributaries provide additional evidence of glaciers. Glaciers typically flow in old stream valleys carving them out from a narrow cross section to a gouged out trough with very steep walls (sometimes called a U-shaped valley). When the ice melts, the glaciated valley form remains in the hard granite and gneiss. However, weak Eocene conglomerate cannot stand at so steep an angle without the support of the ice. The conglomerate then slumps readily into the valley leaving a scarp, poorly drained, hummocky topography, and an abrupt, steep toe. First-rate examples of landslides exist on the north side of the road just northeast of the Lamar River Ranger Station and on the flank of Mount Norris. Trout Lake, north of Soda Butte, fills hollows in the hummocks of a giant landslide that nearly blocked the Soda Butte Creek valley.

The ranger station in the Lamar River valley was built as a ranch to raise bison from about 1907 to the mid-1950s, to preserve the animal from extinction. This valley was also used as a hay field and for food for the bison for several decades during the 1930s to the 1950s. Other historic features in the area now long removed were a small soldier outpost cabin at Soda Butte and a park fish hatchery at

A small landslide with a scarp, hummocks and toe along the road by the Lamar River picnic area.

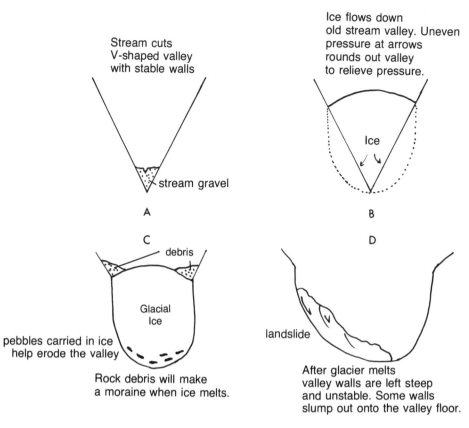

Stream cuts
V-shaped valley
with stable walls

Ice flows down
old stream valley. Uneven
pressure at arrows
rounds out valley
to relieve pressure.

Ice

stream gravel

A

B

C

D

debris

Glacial
Ice

pebbles carried in ice
help erode the valley

landslide

Rock debris will make
a moraine when ice melts.

After glacier melts
valley walls are left steep
and unstable. Some walls
slump out onto the valley floor.

*Sequence of events that have shaped the Lamar River Valley and
many similar valleys in northern Yellowstone.*

Trout Lake.

All dark brown cliffs of Eocene conglomerate contain fossil forests,
(read the separate chapter in the beginning of this book), many
exposed in landslide scarps. Stop at some of the localities shown on
the map to see if you can spot a petrified tree with binoculars. Any
white or light tan spot in the dark brown cliffs is probably a petrified
tree. To reach any of these exposures requires a fairly rugged off-trail
hike—plan to spend at least half a day. However, the hike will be well
worth the effort and time for the fantastic trees and a view of spectacular and unspoiled scenery.

For routes of day hikes read the fossil forest chapter and observe
the routes on the photos of Amethyst Mountain, Specimen Ridge,
Mount Norris, and Mount Hornaday. There are no marked trails so
rugged boots, hiking gear, a helmet for protection from falling rocks
on the steep unstable slopes, a topographic map, and directions from a

Petrified log on Amethyst Mountain.

park ranger are always advisable. Trails in this part of the park may be closed periodically because of bear activity.

No collecting is allowed in the park. If you must have a piece of petrified wood, obtain a Forest Service permit in Gardiner and visit the Tom Miner Basin area outside Yellowstone (see the Livingston—Gardiner roadguide). The Lamar River picnic area provides a nice area to reflect on geology.

Soda Butte, a nearly extinct hot spring vent, is the northeasternmost extension of thermal activity on the Yellowstone Plateau. The canyon walls exposed along Pebble Creek behind the campground are of Madison Limestone. Many fossil remains of shallow-water marine animals that lived here 350 million years ago can be seen in the walls. Ice Box Canyon north of Pebble Creek campgrond is a deep cut through the Eocene volcaniclastic sediments made of gravel from lava of the volcanoes carried here by streams. Cliff exposures along the road show rounded baseball-size chunks of the eroded volcano, and a few pebbles of limestone and granite picked up from the bottom of the valley and included in the flows. The canyon is so named

Volcanic gravel makes cliffs of Mount Hornaday. The best trail to see the fossil trees follows the edge of the vertical cliffs at the very right of the photo. Hummocks in the foreground are a large landslide.

Photo by Utley.

Ice coats the walls of Ice Box Canyon north of Pebble Creek campground well into late spring.

because its steep and shady walls remain coated with ice late into the spring and early summer.

North of Ice Box Canyon, the road enters a small valley with good views of Barronette Peak to the northwest. Almost all of this peak is the dark-brown Lamar River Formation; however, near the base you should be able to see a thin continuous white ledge, one of the Paleozoic limestones with Eocene conglomerate resting directly on top. Also visible, about two-thirds of the way up on the mountain and slightly north of the peak, is a large channel lens of a gray fine-grained ash sediment that presumably formed when ash from a volcanic eruption choked an old stream valley. East of Barronette, pale tan and pink cliffs of Abiathar Mountain of the Cathedral Cliffs Formation become visible. This Formation is the same age as the Lamar River Formation but the material came from a different volcano. It is light-colored because it contains much more quartz than the Lamar River Formation.

From the park entrance to Cooke City are exposures of rocks deposited near the vent complexes of the Eocene volcanoes. Many flows and dark dikes cut across the sediments. The conglomerate and breccia layers here were deposited at an angle on the steep slopes of the volcanoes.

81

Soda Butte, a travertine mound that marks the northern extent of thermal activity in Yellowstone.

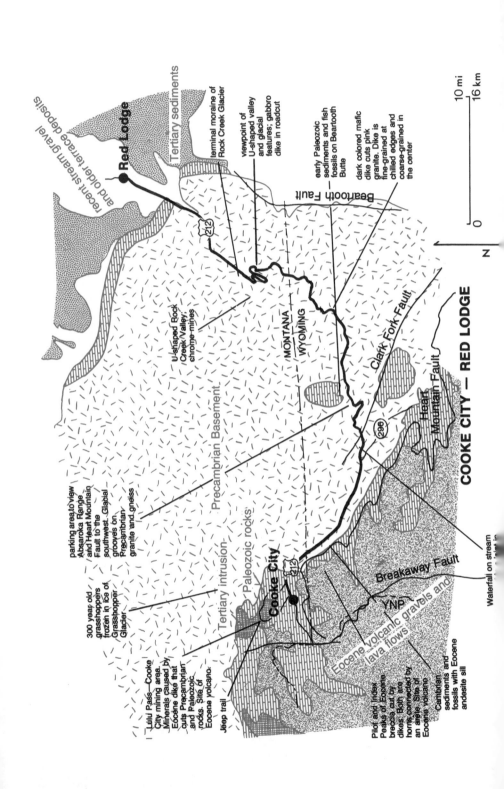

COOKE CITY — RED LODGE

Red Lodge

recent stream gravel and older terrace deposits

Tertiary sediments

terminal moraine of Rock Creek Glacier

viewpoint of U-shaped valley and glacial features; gabbro dike in roadcut

early Paleozoic sediments and fish fossils on Beartooth Butte

dark colored mafic dike cuts pink granite. Dike is fine-grained at chilled edges and coarse-grained in the center

Beartooth Fault

U-shaped Rock Creek Valley; chrome mines

Precambrian Basement

MONTANA
WYOMING

Clark Fork Fault

Heart Mountain Fault

parking area to view Absaroka Range and Heart Mountain Fault to the southwest. Glacial grooves on Precambrian granite and gneiss

300 year old grasshoppers frozen in ice of Grasshopper Glacier

Lulu Pass—Cooke City mining area. Minerals caused by Eocene dike that cuts Precambrian and Paleozoic rocks. Site of Eocene volcano.

Jeep trail

Tertiary intrusion

Paleozoic rocks

Cooke City

Breakaway Fault

YNP

Eocene volcanic gravels and lava flows

Waterfall on stream

Pilot and Index Peaks of Eocene breccia cut by dikes. Both are horns connected by an arête. Site of Eocene volcano.

Cambrian sediments and fossils with Eocene andesite sill

N

0   10 mi
0   16 km

Cooke City—Red Lodge
65 mi./105 km.

The road from Cooke City to Red Lodge crosses extremely diverse rocks and is one of the most beautiful roads in the United States. The Cooke City area is near a complex of Eocene volcanoes that produced much of the debris that now buries the fossil forests. Cliffs above the city show numerous thin black vertical dikes, the plumbing of fractures the magma followed as it rose to the surface through the breccias and conglomerates.

Republic Mountain visible to the south from Cooke City, is the breakaway point for the Heart Mountain thrust fault. According to W. G. Pierce of the U.S. Geological Survey, a huge slab of the Madison Limestone broke away from this point and slid southeast to the area north of Cody, where it now lies on top of much younger Eocene rocks. Right after the limestone slid, the Absaroka volcanoes erupted and covered the fault plane with debris. The original slab may have been about 500 square miles but then broke up into a number of blocks that are scattered over 1300 square miles between Cooke City and Cody. Some of the blocks may have moved over 50 miles.

Just north of Cooke City these Eocene dikes and other rocks that formed in the throat of the volcano cut through (intrude) Precambrian metamorphic rocks and Paleozoic sedimentary rocks. During the intrusion, hot water altered the surrounding country rock, as it is now doing in the Grand Canyon of the Yellowstone (see the Roosevelt—Canyon roadguide). As part of the alteration, many ore minerals were precipitated from the hot water. These minerals have been mined off-and-on since the late 1800s, when Cooke City was founded as a small mining town. The general store in Cooke City displays a scrapbook of photos taken in the area around the turn of the century showing various mines. Economically important commodities that formed by a combination of a heating of the country rock and hot water alteration include: gold, copper, lead, silver, and zinc. The gold and copper have made the most money for the miners.

A small dirt road just north of Cooke City leads up to Lulu Pass and into the mining area. If you have four-wheel drive or other vehicle suitable for rugged roads, this is a good side trip. It is usually best to ask someone in town about the road conditions before you start because snow remains well into the early summer. Crown Butte, at the top of the pass, is made of Cambrian sediments cut by a large dike, which was the focus of considerable mining activity.

Notice the long linear strips up the sides of the mountain where all of the trees are stripped off. The trees were destroyed by snow avalanches in the winter. About three miles from here along a rugged jeep or hiking trail is Grasshopper Glacier. About 300 years ago a swarm of grasshoppers was blown onto the ice, buried by winter snows and incorporated into the glacier. Now they have flowed through the glacier and are melting out at the toe.

Snow avalanche paths near Lulu Pass north of Cooke City.

Northeast of Cooke City, the main highway (U.S. 212) winds up over Colter Pass. One large outcrop south of the road 2.5 miles southeast of Colter Pass and almost on the Montana/Wyoming state line exposes a sill of andesite lava sandwiched between layers of Paleozoic marine sedimentary rock of the Cambrian Park Shale. Sills form as liquid magma injects between layers of sedimentary rock.

Exposures along the road near the base of Pilot and Index peaks. The white cliffs are lower Paleozoic Limestones. The Upper Paleozoic section slid east along the Heart Mountain Thrust fault. The upper dark cliffs are Eocene volcanic conglomerates that buried the fault.

Dikes, on the other hand, cut the bedding at an angle. This 50 million-year-old Eocene sill was intruded during volcanic eruptions which were associated with burial of the fossil forests. Pilot and Index peaks are near the sites of the old eroded volcanoes. Binoculars reveal many dark strips of dikes and sills on the sides of the mountain.

Three miles southeast of the state line, is an outcrop of Cambrian Flathead Sandstone to the south of the road. The sandstone rests on Precambrian basement and shows many burrows and trails of marine invertebrates.

Pilot and Index peaks look very much like the Matterhorn in Switzerland and for exactly the same reason—all three were carved by glaciers. Some interesting things happen at the base of the glacier. Water expands when frozen, but if placed under pressure will melt. The pressure at the base of many glaciers is great enough to cause a thin layer of the ice to melt and flow into any small cracks in the rock where it instantly refreezes. When the water refreezes, it again

85

Pilot and Index peaks—two glacial horns connected by an arête ridge.

expands, breaks out a chunk of rock, and includes it in the glaciel ice. In this manner, glaciers erode the sides of mountains into steep-cliffed amphitheaters called cirques. If three or four glaciers occur on all sides of a mountain, they meet leaving a tall pinnacle called a horn. Where two glaciers almost meet, they leave a long knife-like ridge called an arête between them; the ridge between Pilot and Index peaks is a good example. The Bear's Tooth visible from the top of Beartooth Pass is another fine example of a glacial horn. Watch for the numerous examples of all of these features throughout the Bear-tooth Plateau.

Glacial grooves in Precambrian granite along Montana 212 just north of the junction with Montana 296.

As the glacial ice drags along plucked particles of rock, they gouge grooves into the bedrock. A good place to see such grooves is about a quarter of a mile north of the scenic viewpoint at the top of the hill north of the junction with Wyoming 296.

Between Colter Pass and the junction with Wyoming 296, the road follows a major fault. To the north, Precambrian basement rocks have been uplifted as a giant block several thousand feet; these same rocks

Beartooth Butte, an erosional remnant of Lower Paleozoic sedimentary rock on top of the Beartooth uplift. The butte is made of Cambrian rocks near the base, Ordovician and Devonian rocks near the top. The large dark spot is the cross section of a channel of Devonian siltstone, which contains plant and fish fossils. The hummocky ground between the lake and the cliffs is a large landslide.

exist that far beneath the surface on the south side of the road, but are covered by younger sediments. All of the Paleozoic sediments as well as the Eocene conglomerates have slid or been eroded off the Beartooth block leaving naked Precambrian granite, gneiss, and various dikes ranging in age from Precambrian to Cretaceous. About 7 miles southeast of Colter Pass, where the road crosses the Yellowstone River, you can also see evidence of this Heart Mountain fault south of the road. There, Eocene conglomerate rests on a prominent white layer of middle Cambrian Pilgrim Limestone; the upper Paleozoic was thrust to the southeast.

Stop at the small un-named waterfall just north of the junction with Wyoming 296. A short walk to the abandoned bridge allows the best view of the falls. Notice that the entire course of the stream is whitewater as it drains the top of the Beartooth Plateau. The stream rushes down a straight line under the old bridge because it follows a joint or crack in the bedrock.

Beartooth Butte provides evidence that Paleozoic rock once covered the Precambrian crystalline rock. The butte is a remnant of Paleozoic sediments that were not quite eroded away, or were left behind when the rest of the Heart Mountain plate slid off to the east. Starting at the level of Beartooth Lake, these sediments include Cambrian sandstone, Ordovician, and Devonian marine rocks. The butte is famous for Devonian fossil fish that occur near the summit. The easiest place to find them is in the red shales that have tumbled down onto the talus cone at the base of the cliff. The large lens of red siltstone in the center of the butte is the cross-section of an old channel.

Continuing north and east from Beartooth Butte, the road crosses a high flat meadow that supports a scrubby growth of tree-line spruce trees. From here to the top, the road crosses true tundra with permanent frost not far beneath the surface of the ground. Large polygons and circles in the ground above the tree line are caused by frost action. This patterned ground is typical of tundras in arctic latitudes as well as those at high elevations. Just past the Top-of-the-World Cafe, the road starts a tortuous climb to the top of the highest pass in the United States that is not on the Continental Divide. This road can be closed for severe snow storms just about any time during the summer and is, of course, closed all winter. A few permanent snow fields and a possible scrappy remnant glacier or two remain all summer. Algae that grow in snow color some of the permanent snow fields red during the summer.

Most of the rock along the switchbacks to the summit is light pink granite and gneiss, along with a few black dikes up to several hundred feet thick that were intruded into the granite. All of the dikes are basaltic, and some show a chilled margin. The rock at the edge of the dike is extremely fine-grained, with crystals too small to see because it cooled quickly in contact with the cold granite. However, the center of the dikes cooled slowly and formed diabase or gabbro with crystals large enough to see without a magnifying lens.

North of the second pass and just into Montana there is a viewpoint/rest area that is worth a visit and a short walk to see many fine glacial features. To the west is the main part of the Beartooth Uplift—all now above the tree line. Glaciers gouged Rock Creek valley into a broad, straight steep-walled trough. A small winding

Glacially carved basement rock on the Beartooth Plateau.
Photo by Dave Alt.

View up the Rock Creek U-shaped glacially carved valley.
Photo by Dave Alt.

dirt road leads up the sheer side of the valley to one of the now abandoned mining areas where chromite, concentrated by a Precambrian intrusion, was mined during World War II. A walk up or down the road from the parking area will allow you to see most of the rocks that form the Precambrian basement of the Beartooth Plateau.

The road from the view area to Red Lodge winds down into the glaciated steep-walled trough of Rock Creek valley, and follows it nearly to town. The road crosses a large terminal moraine about twelve miles from Red Lodge that marks where the glacier stopped. The last outcrops of Precambrian granite are about seven miles south of town. From there on, the road crosses Paleozoic and Mesozoic sedimentary rocks that were eroded off the top of the Beartooths. The thick prominent white layer is the Madison Limestone that formed as a reef in a shallow sea.

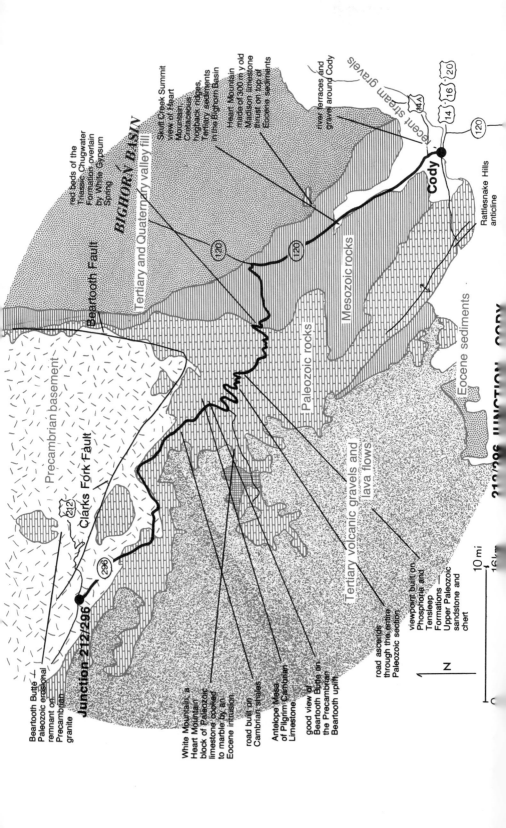

BIGHORN BASIN

red beds of the Triassic Chugwater Formation overlain by White Gypsum Spring

Beartooth Fault

Precambrian basement

Clarks Fork Fault

Beartooth Butte — Paleozoic erosional remnant on Precambrian granite

Junction 212/296

White Mountain, a Heart Mountain block of Paleozoic limestone cooked to marble by an Eocene intrusion

road built on Cambrian shales

Antelope Mesa of Pilgrim Cambrian Limestone

good view of Beartooth Butte on the Precambrian Beartooth uplift

Sheet Creek Summit — view of Heart Mountain

Cretaceous hogback ridges, Tertiary sediments in the Bighorn Basin

Heart Mountain — mass of 300 m y old Madison limestone thrust on top of Eocene sediments

river terraces and gravel around Cody

Tertiary and Quaternary valley fill

Paleozoic rocks

Mesozoic rocks

Eocene sediments

Tertiary volcanic gravels and lava flows

viewpoint built on Phosphoria and Tensleep Formations — Upper Paleozoic sandstone and chert

road ascends through the entire Paleozoic section

Cody

Rattlesnake Hills anticline

recent stream gravels

N

10 mi

16 km

212/296 JUNCTION, CODY

Junction of U.S. 212 and Wyoming 296—
Cody (Dead Indian Pass)
70 mi./114 km.

The road leading east from the junction of U.S. 212 and Wyoming 296 is a scenic drive across exciting rocks. The gravel road over the summit is passable to any car in good condition if caution is used on the sharp switchbacks. Not long ago the entire road was gravel, but more is paved every year.

For the first few miles, roadcuts expose Precambrian granitic gneiss of the Beartooth basement complex. Numerous scenic pullouts provide places to stop and look at the pink potassium feldspar-rich rock. One good place is a roadcut 2.4 miles east of the junction of U.S. 212 and Wyoming 296. A pegmatite vein on the west edge of the cliff exposes some very large quartz and feldspar crystals. This stop also furnishes good views back west to Pilot and Index peaks, made of intrusive and mudflow rocks of the Eocene Absaroka volcanics and later carved to their present shape by glaciers.

From the highway junction to Dead Indian Summit, the road and the Clark Fork River follow two major faults. To the northeast, Precambrian rocks were pushed up along the Clark Fork fault just on the other side of the river. Notice that even from a distance, by matching similar landscapes, you can tell that the same granite at road level here is several thousand feet higher on top of the Beartooth Mountains.

Slabs of Paleozoic sedimentary rock tilted up on edge by the uplift of Precambrian basement of the Beartooth block to the north. Photo is looking west from Wyoming 120 just north of the junction with Wyoming 296.   Photo by Dave Alt.

On the southwest side of the road, the Heart Mountain thrust fault winds along the base of the Absaroka mountain chain. W. G. Pierce of the U.S. Geological Survey has suggested that sometime during Eocene time, Paleozoic limestones broke away from an area around Republic Mountain near Cooke City, and moved southeast. The rocks may have broken because the entire area started to swell from expansion caused by molten rock moving to the surface at the beginning of the Absaroka volcanism, or from uplift of the Beartooth block. The layered sedimentary rock slid like the top card on a steeply tilted deck. This layer slid along the fault plane riding up over itself, then over the ground surface across Eocene rocks north of Cody. There, at Heart Mountain, 300 million-year-old Paleozoic limestone (Madison Limestone) now lies on 55 million-year-old volcanic ash of the Eocene Willwood Formation! We know the fault moved sometime after the Eocene rocks were deposited. Immediately after the limestone slid, the Absaroka volcanoes erupted, covering the fault plane with volcanic flows and conglomerate. Geologists can date these volcanic rocks both by the plant fossils they contain and by radiometric methods. The fault plane can now be seen south of the road by following the prominent white ledge of Cambrian limestone in contact with overlying Eocene volcanic gravel; all of the middle-Upper Paleozoic rocks were thrust southeast.

If you look south from along the Clark Fork River, you can see a thin white ledge or band of rocks that continues along the whole valley wall, and is covered by brown volcanic conglomerate. This is the actual trace of the fault. The white ledge is lower Paleozoic limestone (Cambrian Pilgrim Limestone) and the brown rocks are volcanic gravels and lava flows of the Absaroka group. In many places, both the Paleozoic rocks and the conglomerates are cut by dikes that fed magma to the volcanoes. Here and there, chunks of Upper Paleozoic limestone broke off the main slab and were left behind. Some have now been left as small peaks, and others can be identified by a sudden thickening of the white Paleozoic sedimentary rocks.

Just south of the Clark Fork River crossing, the road crosses into early Paleozoic sedimentary rocks. Parts of this road have had problems with landslides and slumps because it was built on slick unstable Cambrian shales. Just as the road starts to descend into Sunlight Creek canyon there is a turnout on the north side, a good place to look at the scenery and think about geology.

At this point the Clark Fork fault turns north and becomes the Beartooth fault, which raised the Beartooth Plateau on the east side. The Precambrian rocks have been shoved up over 20,000 feet along the Beartooth fault. That is, the same rocks on the top of the plateau exist 20,000 feet lower beneath Tertiary, Mesozoic and Paleozoic rocks of the Bighorn Basin to the east. You can tell that Paleozoic rocks once covered the granites of the plateau by looking at Beartooth Butte, a flat-topped mountain on the northern skyline. The butte is early Paleozoic sediments that were left after all the others either eroded or slid away. If you look closly, you can see how they are correlated with the nearly vertical Paleozoic layers east of the Clark Fork River that were turned up on the east side of the fault.

The road here at the start into the canyon is built on Cambrian sediments covered in places by glacial moraines. Back to the northwest on the north side of the road you can see Antelope Mesa, the mountain that looks like a large elevated golf green. This is one of the blocks of the Heart Mountain thrust, as is Sugarloaf Mountain to the west and on the south side of the road. At the base of the canyon, a dirt

Paleozoic sedimentary strata exposed on the cliffs at the upper right lie on Precambrian crystalline basement uplifted along the Beartooth fault in the canyon in the center of the photo.

road leads into Sunlight Basin, an interesting side trip to see more of the Heart Mountain thrust fault and cliffs of the Absaroka volcanics. From the base of the canyon to the top of Dead Indian Pass, the road winds around many switchbacks that expose the entire Paleozoic section from Cambrian sandstone to Permian rocks at the top. Consult the stratigraphic section here if you are interested in the names of the various formations. Starting at the contact with the Precambrian basement, you should be able to recognize: Flathead Sandstone, Cambrian shales, Ordovician dolomite, Devonian shale, Madison Limestone, Tensleep Sandstone, and the Phosphoria Formation. The viewpoint at the top is on quartz sandstone of the Tensleep Formation. Scattered edges of the Permian Phosphoria Formation exist above. The best viewpoint can be reached by a short hike to the northwest along the ridge above the parking area.

The view west is a fine geologic panorama. First, look down the Clark Fork Valley with Precambrian rocks of the Beartooth Uplift on the north and Eocene volcanic conglomerates resting on Paleozoic sedimentary rocks to the south. The contact between them is the trace of the Heart Mountain fault. In the distance you should be able to see Beartooth Butte at the skyline.

The floor of Sunlight Basin exposes Paleozoic sedimentary rocks, and the base of the hills at the edge of the valley is the contact with the Heart Mountain thrust. White Mountain is a block of limestone left behind as the main slab slid east. The limestone has now been cooked to marble by the heat of some dikes associated with one of the Eocene volcanoes. Several glacial moraines can be seen at the base of Steamboat Hill, another block from the Heart Mountain fault.

Farther east, the road crosses Tensleep Sandstone, made of well-sorted medium-grained quartz sand deposited in a large windblown coastal dune field. Some good exposures of cross-beds that form as the dunes move appear on the far wall of a small canyon a mile or so east of the summit. As the road starts to switch back down into the valley, you are driving on the bedding surface of the Phosphoria Formation. In other areas of the Yellowstone Country, such as around Pocatello, Idaho, this formation is mined for the phosphate for fertilizer and explosives. Stop at one of the turnouts to look downhill at the bright red beds of the Triassic Chugwater Formation at the base of the hill.

The Chugwater Formation is the oldest Mesozoic formation in this region, an excellent marker bed throughout Wyoming because it is red. The color comes from iron stains on mineral grains. Many Triassic formations throughout the world are red, for reasons that no one clearly understands. Overlying the red beds is a white layer of the Gypsum Springs Formation that contains various evaporite minerals that precipitated out of solution in a desert lake environment.

From the red outcrops of the Chugwater Formation to Wyoming

Red beds of sandstone and siltstone of the Triassic Chugwater Formation overlain at the top by white evaporite deposits of the Gypsum Springs Formation. Grassy slope along the road at the lower left is the dip slope of the top of the Phosphoria Formation.

120, the road crosses Cretaceous sands and shales. Many of the rock layers have been tilted up on edge by the Beartooth Uplift, and make long hogback ridges of formations that lie deeply buried in the Bighorn Basin. Many of the buried formations produce considerable oil and natural gas from organic matter in the original sediments. In the distance, and straight ahead to the east, you can see Heart Mountain. The massive cliffs at the top of the mountain are 300 million-year-old Madison Limestone transported here from the vicinity of Cooke City! The base of the mountain is made of flat-lying Tertiary sediments of the 55 million-year-old Willwood Formation. At the junction with Wyoming 120, turn right to Cody.

All rocks from Wyoming 120 until the bridge crosses the river at the edge of town are Cretaceous sandstones and shales. The top of Skull Creek Summit is a good place to stop and look back to the north along the edge of the Bighorn Basin. To the west are Mesozoic and Paleozoic sedimentary rocks tilted up on edge against the 20,000-foot uplift of Precambrian basement that makes up the Beartooth Plateau. These sediments are flat-lying and deeply buried out in the basin. All of the rocks in the basin are Tertiary except for an occasional block of Paleozoic limestone that was thrust over them. About 12 miles north of Cody and five miles south of the turn onto Wyoming 120, you can see a small syncline (west) and anticline (east) in the Cretaceous rocks by looking north.

A few miles north of Cody, a quarry on the left exposes the Cretaceous Mowrey Shale and a white bed of weathered volcanic ash called bentonite. Bentonite is actually a kind of clay mineral left after chemical alteration of the ash. When wet, the clay takes water into its molecular structure, thus swelling to make a very slippery surface. The town of Cody is built on stream gravel washed down the Shoshone River from the Absaroka Range. Some of the gravels can be seen as you cross the river into town.

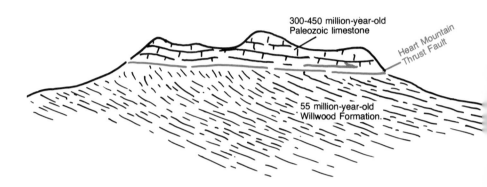

Heart Mountain north of Cody is made of a chunk of Madison and older limestone from the vicinity of Cooke City thrust over Eocene sediments of the Willwood Formation.

Tower Junction—Canyon
19 mi./31 km.

From Tower Junction to Canyon, the road crosses mostly Eocene volcanic gravels and lava flows, largely covered by Quaternary glacial and stream gravels, as well as ash and lava flows of the Yellowstone Plateau volcanic field. Here and there, warm water heated by the hot rock not far below the surface seeps up through the older rocks to make a small hot spring. Just north of Canyon, the road crosses onto the Yellowstone caldera, a large basin now filled with thick rhyolite lava flows to make a nearly flat high plateau. Most of the volcanic rocks exposed within the caldera formed from viscous rhyolite lava during the last 600,000 years.

The canyon of the Yellowstone River at Calcite Springs in the vicinity of Tower Falls exposes dark brown to greenish conglomerate of the Absaroka volcanic pile, the same rock that buries petrified trees throughout northwestern Yellowstone. Near the base of the canyon, several steam and hot water vents exit through the rock to make small steam clouds on cool days. Near the vents, the surrounding rock has been altered by chemicals in the hot water to bright yellows and oranges.

Several good views of the stratigraphy in the canyon walls can be seen on the far side of the canyon from any of the pull-outs along the hill north of the Tower Falls store. Probably the best viewpoint, and one with some interpretive signs, is along a short trail from a parking area for Calcite Springs overlook. About two-thirds of the way up the cliffs, the Eocene conglomerates are overlain by a series of basalt flows and stream gravels. The basalt flows contain outstanding examples of columns defined by shrinkage cracks that formed as the lava solidified. These basalts, called the basalt of the Narrows, are about 1.5 million years old, and came from lava that poured out of vents to the south on the Yellowstone Plateau volcanic field. The

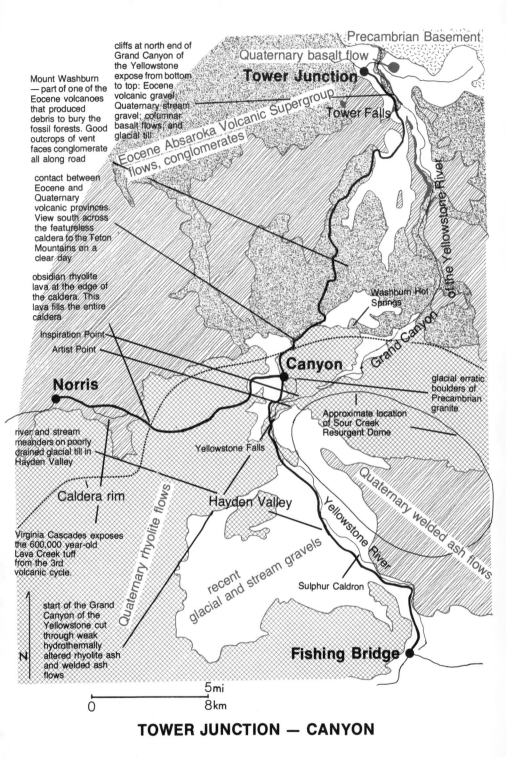

Mount Washburn — part of one of the Eocene volcanoes that produced debris to bury the fossil forests. Good outcrops of vent faces conglomerate all along road

cliffs at north end of Grand Canyon of the Yellowstone expose from bottom to top: Eocene volcanic gravel; Quaternary stream gravel; columnar basalt flows; and glacial till

Precambrian Basement

Quaternary basalt flow

Tower Junction

Eocene Absaroka Volcanic Supergroup flows, conglomerates

Tower Falls

Grand Canyon of the Yellowstone River

contact between Eocene and Quaternary volcanic provinces. View south across the featureless caldera to the Teton Mountains on a clear day

obsidian rhyolite lava at the edge of the caldera. This lava fills the entire caldera

Inspiration Point

Artist Point

Washburn Hot Springs

Canyon

Grand Canyon

Norris

glacial erratic boulders of Precambrian granite

Approximate location of Sour Creek Resurgent Dome

river and stream meanders on poorly drained glacial till in Hayden Valley

Yellowstone Falls

Quaternary welded ash flows

Caldera rim

Quaternary rhyolite flows

Hayden Valley

Yellowstone River

Virginia Cascades exposes the 600,000 year-old Lava Creek tuff from the 3rd volcanic cycle.

recent glacial and stream gravels

Sulphur Caldron

N

start of the Grand Canyon of the Yellowstone cut through weak hydrothermally altered rhyolite ash and welded ash flows

Fishing Bridge

5mi

0   8km

TOWER JUNCTION — CANYON

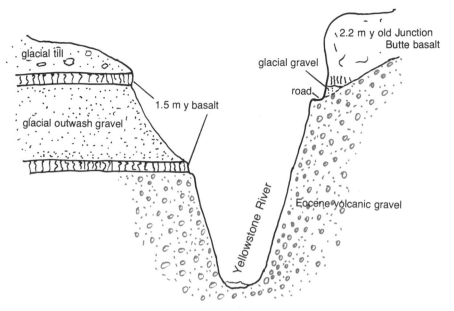

glacial till

1.5 m y basalt

glacial outwash gravel

glacial gravel

road

2.2 m y old Junction Butte basalt

Yellowstone River

Eocene volcanic gravel

Generalized cross section across the Yellowstone Canyon as seen from Calcite Springs Overlook at the Narrows. Road at the overhanging cliffs of Junction Butte Basalt just north (uphill) of Tower Store indicated by the arrow. Redrafted and modified from Field (1932).

stream gravels between the two conspicuous flows on the north side of the canyon are probably glacial outwash. Poorly sorted glacial till caps the rim of the canyon.

Several pull-outs along the road about a quarter of a mile north of the Tower Falls store provide close views of a basalt flow. The thick basalt flow along the road is the Junction Butte Basalt and was deposited in an ancient stream channel. This basalt is about 2 million

Cliffs at view area west of Tower Falls. Dark vertical cliffs at the base are Eocene volcanic gravel overlain by slopes of lighter-colored Pleistocene stream and glacial gravel. Thin cliff near the top is a columnar jointed basalt flow overlain by glacial till.

A flow of 2.2 million-year-old Junction Butte Basalt along the road between Calcite Springs overlook and Tower Falls.

years old—.5 million years older than either of the two columnar jointed Narrows Basalt flows on the other side of the canyon. Notice that the basalt rests on a thin layer of gravel. If you walk down the road, you can see outcrops of the dark brown Eocene conglomerate.

Tower Falls flows over a vertical cliff in the Eocene conglomerate. A large boulder perched precariously at the very lip of the cascade looks like it should tumble any minute. If you think of betting on when it will fall, remember that members of the Hayden survey of 1871 cast bets on the hour when this same boulder would fall! Just upstream from the falls are some tall thin obelisks with a large boulder capping the top. These are called hoodoos, and they form

Tower Falls and cliffs of Eocene volcanic gravel.

because the boulder protects the weak conglomerate from rainsplash erosion.

South of Tower Falls, the road climbs up the flank of Mount Washburn, one of the volcanoes that erupted 50 million years ago to produce the mudflows that buried the fossil forests. Along a hairpin turn near the top of the mountain are some angular conglomerate beds dipping at about 30 degrees. These are the original dips of rock deposited on the slopes of the old volcano. Numerous outcrops of the angular conglomerate debris exist where the road levels off near the summit of Dunraven Pass. Pull-outs along the highway, or a side trip (hike) on the Chittenden road part way up Mount Washburn, provide spectacular views into the valley.

View from the south of Dunraven Pass southwest across the edge of the Yellowstone caldera marked by the low hills near the top of the photo.

South of Dunraven Pass, the landscape falls away to provide a spectacular view to the south. This is the very edge of the giant Yellowstone caldera from the third eruptive cycle of 600,000 years ago that erupted many of the Quaternary lava flows throughout the Yellowstone Country. On a clear day, you can see low hills 35 miles way on the far rim of the caldera. The crater is not a deep hole in the ground as with some volcanoes, like Crater Lake, Oregon, because after the caldera formed, it filled with viscous rhyolite lava. When molten, this lava was quite stiff and piled up like bread dough to thicknesses of nearly 1000 feet as it flowed. The way that rhyolite flows to produce thin continuous layers, like the Huckleberry Ridge Tuff on Mount Everts, is for the lava to shatter by a steam explosion and for a cloud of the tiny particles mixed with ash to flow as a cloud.

The Lower Falls of the Yellowstone River. Here, the river has cut deep into soft rhyolite chemically altered by hot water.

This cloud can remain so hot that the particles fuse together into solid rock called a welded-ash flow.

As the road descends from Dunraven Pass you can get a good view of the Washburn Hot Springs downhill to the east. There must be a limited supply of ground water here because many of these springs are dominated by vapors. Apparently most of the ground water boils off as steam before more water can flow in to take its place. About half a mile south of the Washburn Hot Springs and three miles north of Canyon Junction, the road leaves the Absaroka volcanics (the Sepulcher Formation here) and crosses onto the plateau rhyolite flows.

At Canyon, the Yellowstone River has cut deep into some of the rhyolite lavas. Reactions between rock and steam create the bright yellow, orange and red colors. The water soaks deep into the earth from rain and snow-melt at the surface, heats in contact with partially molten rocks that may still exist only one and a half to three miles below this area, and returns to the surface through hot springs. The hot water also dissolves minerals from the hot rocks, and deposits them in the cooler rocks near the surface; and around hot springs to chemically alter the ryolite lava into the bright colors. Such chemically altered rhyolite is very weak and erodes easily, allowing the river to carve the deep canyon. At the Upper Falls, notice that the rock is dark brown unaltered rhyolite. This rock resists erosion and forms the nick point for the falls that defines the upstream extent of the canyon.

Plan to stop at the Canyon area for a short visit. There is a fine visitor center that includes a discussion of the geologic processes that are forming the canyon. The one-way drive from Canyon Junction along the north rim of the canyon to Inspiration, Grandview, and Lookout points shows many of the features of the canyon. However, to fully appreciate the geology, plan also to visit the Upper Falls and then cross to the other side of the canyon to Artist Point (see the Canyon—Fishing Bridge roadguide).

Large granite boulders are scattered through the trees around the rim of the canyon. Because there is no granite bedrock anywhere near the canyon, these must have been brought here by glaciers, and are called glacial erratic boulders. The nearest granitic bedrock is the Beartooth Mountains about 40 miles to the north, so the boulders must have been carried from at least that far away. It is interesting to speculate how it happens that both sides of the canyon and the entire plateau surface were glaciated and yet the canyon is not a steep-walled glacial trough and shows no other signs of having been carved by glaciers. Because pre-Pinedale age sediments occur in the canyon, geologists know that it existed before it was covered by ice. The canyon may have kept its steep walls because stagnant ice blocked the canyon to trap sediment and because the ice here was an ice cap covering the entire plateau, rather than a mountain valley glacier.

NORRIS — CANYON

See the map on page 98.

Norris—Canyon
12 mi./19 km.

The route from Norris Junction to Canyon crosses Quaternary rhyolite erupted from the Yellowstone caldera now covered in a few places by glacial and stream gravel and hydrothermally altered ground from hot springs and geysers. From the junction 1.6 miles east to the turn-off for the Virginia Cascades drive, the road crosses Pinedale glacial till and stream sediments. The side road to Virginia Cascades passes good exposures of the youngest layer of welded-ash, called the Lava Creek Tuff, produced during the last (third) major explosive cycle that made the Yellowstone caldera about 600,000 years ago. This welded-ash lies just outside the rim of the caldera, and is exposed here through erosion by the Gibbon River.

The road crosses the rim into the caldera, about four miles after leaving Norris, at the bottom of the hill just after the road crosses the Gibbon River and east of the exit of the one-way Virginia Cascades drive. The caldera wall here is buried under some of the thick rhyolite flows that later filled the caldera after it collapsed. The featureless caldera to the south and east is now covered with dense lodgepole pine forests. Most rocks here are rhyolitic lava flows that oozed out and filled the caldera long after it exploded and collapsed. One of these flows, exposed about a half mile east of the exit of the Virginia Cascades drive, has been dated by the radioactive decay of potassium to argon to be about 100,000 years old. From the top of the hill there is a view to the north and west of uplifted Paleozoic rocks that makes the mountains of the southern end of the Gallatin Range.

Remember that the earth's crust here is extremely thin and very hot. Rock is still cooling from the geologically recent eruptions. Ongoing uplift not far to the south may reflect entry of molten rock only a mile or two below the surface. All of the geysers, hot springs, mud volcanoes, and steam vents are evidence of hot rock at extremely shallow depth.

Some volcanic rocks altered and weakened by hydrothermal (hot water) alteration occur about four miles east of the top of the hill. Cascade Meadows are glacial outwash and stream sand and gravel covering volcanic rhyolite bedrock. Canyon Junction is at the east edge of the meadow. There are some very good displays at the Canyon Visitor Center that explain some of the local geology and the origin of the canyon. Read the next section for a discussion of the geology exposed along the Grand Canyon of the Yellowstone River.

Canyon—Fishing Bridge
16 mi./26 km.

South of Canyon Junction, the road is built on Pinedale glacial outwash sand and till that cover various rhyolite lava flows. About two miles south of the junction, a short side road leads to the rim of the Upper Falls. A short trail goes to the very lip of the cascade, where you can get a good look at the unaltered rhyolite. This is also a good place to see the ongoing process of headward stream erosion that is carving the entire canyon. Erosion is probably proceeding at a slower rate today than in the past because the river has finally eroded through the weakened hydrothermally altered rock and Eocene conglomerate to the north and into hard unaltered rhyolite.

Just south of the road to the Upper Falls, and 2.5 miles south of Canyon Junction, another side road leads to viewpoints on the south rim of the canyon. These provide some of the most spectacular views of the rocks. The oldest known sinter deposits in the park are exposed along the road on the south side of the canyon. Artist Point is an excellent place to observe the beautiful colors produced in the rhyolite walls of the canyon from hot water alteration and chemical reaction. Notice the tall rock spires and hoodoos eroded in the soft canyon walls.

The Upper Falls of the Yellowstone flow over resistant unaltered rhyolite.

CANYON — FISHING BRIDGE

See the map on page 98.

The road between the Upper Falls and Fishing Bridge passes few good outcrops. Those few rocks that you can see are Quaternary rhyolite ash and lava flows, glacial gravel, recent stream and lake sediments, and ground altered by hot water around the hot springs.

Hayden Valley, 4.5 miles south of Canyon, was once filled by an arm of Yellowstone Lake and contains fine-grained lake sediments made of clay, silt, and sand now covered with glacial till. Because the glacial till contains all different grain sizes, including clay to plug the pore spaces, and the veneer of lake sediments, water cannot soak into the ground in this area. That is why Hayden Valley is so swampy and the stream produces interesting and tortuous meanders as it lazily flows across the wet valley.

About 5.5 miles south of Canyon Junction and a mile or so into Hayden Valley, the road climbs onto a small hill of the 100,000-year-old Hayden Valley rhyolite flow that is covered by some Pinedale glacial till, older Bull Lake glacial outwash sediments, and stream gravel. Just over half a mile farther south there is a pull-out for a view of Hayden Valley and some other interesting geologic features. The mountain range to the east is the Absaroka Range made of Eocene volcanic material. A resurgent dome in the eastern part of the Yellowstone caldera, called the Sour Creek Dome, is visible to the immediate southeast. The Sour Creek Dome probably formed soon after the caldera-forming phase of the third volcanic cycle 600,000 years ago, and is now extinct. Renewed uplift along a football-shaped area that includes much of central Yellowstone is bulging upwards from the pressure of underlying magma at almost an inch per year. Even though this may sound slow, it is a tremendously fast uplift rate seen in the perspective of geologic time and is faster than your fingernails grow. If this uplift occurs long enough, it could produce a dome or even be the precurser to additional volcanic activity.

The Trout Creek pull-off provides a view of an interesting double meander and some cliffs of lake sediment. One of the cliffs above the

An interesting meander of Trout Creek in the Hayden Valley. Cliffs are made of lake sediments deposited when an arm of Yellowstone Lake once filled Hayden Valley.

stream has an ancient channel of stream gravel. Apparently, there were at least two lakes in Hayden Valley. The lakes may have formed when glacial ice dammed the streams or from elevation changes caused by the recent uplift.

Sulphur Cauldron and Mud Volcano, about half way between Canyon and Fishing Bridge, are also worth a stop. The thermal features here are mostly mud pots and fumaroles because the area does not have enough ground water to make geysers or hot springs. The fumaroles occur where the ground water boils away faster than it can be recharged. The vapors are rich in sulfuric acid that leaches the rock, breaking it down into clay. Because no water washes away the acid or leached rock, it remains as sticky clay to form a mud pot.

Earthquakes in 1978 changed the plumbing of this thermal system. New pots have developed, killing some trees, and other pots have snuffed out. The earthquakes are caused by continuing uplift of the area. This is one location where it is especially important to stay on the board walks. The soil made from altered clays around some of the thermal areas forms a very thin crust that looks like solid ground but will not hold the weight of a person. Many people have been needlessly scalded here after straying from a safe walk and breaking through what they thought was solid ground.

Between Mud Volcano and Lake Junction, the road follows the Yellowstone River. Here the river meanders within a notch where it downcut into the Quaternary lavas. The Le Hardy Rapids exist where the river crosses a fault. Submerged terraces in Yellowstone Lake show that the lake once stood at a lower level—either due to uplift and tilting of the ground or because there is now more water in the lake.

Madison—West Thumb
33 mi./53 km.

The road from Madison to Old Faithful, West Thumb and then on to Fishing Bridge is entirely within the Yellowstone caldera that resulted from the collapse of the magma chamber during the last explosive, caldera-forming volcanic eruption 600,000 years ago. The caldera has since filled with lava from intermittent flows produced between 600,000 and about 70,000 years ago. Geologic features along this route include the lava flows, geysers, and hot springs associated with the still cooling magma, sinter deposits from thermal features, hydrothermal explosion craters, an extinct resurgent dome, and a skim of glacial sand and gravel covering the lava flows.

The road along this route crosses through four of the major geyser basins in the park. These include the Lower, Midway and Upper geyser basins and the West Thumb Basin. Two additional interesting thermal areas, Lone Star and Shoshone, are in the backcountry south of the road. If you enjoy geyser gazing in a completely unspoiled area, it is well worth the three to eight mile hike to these basins.

Firehole Falls along the one-way southbound side road south of Madison Junction exposes one of the giant rhyolite lava flows. Rhyolite usually flows sluggishly because its lava is very stiff. However, some of that in the caldera actually did flow distances of up to 18 miles. In places, such as Madison Canyon, 9-12 miles from the vent, the lava ponded to a thickness of 500 to 1000 feet. Often the tops of rhyolite flows cool and harden while the center remains molten, and continues to move. this causes the hardened surface to crack into a pile of rubble, called breccia. Cliffs along the Firehole River drive expose just such a thick rhyolite flow with a rubbly broken breccia zone at its top near the top of the falls.

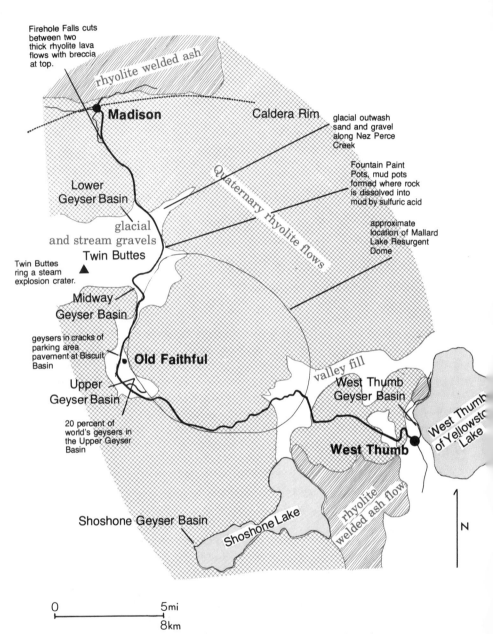

Firehole Falls cuts between two thick rhyolite lava flows with breccia at top.

rhyolite welded ash

Madison

Caldera Rim

glacial outwash sand and gravel along Nez Perce Creek

Lower Geyser Basin

Quaternary rhyolite flows

Fountain Paint Pots, mud pots formed where rock is dissolved into mud by sulfuric acid

glacial and stream gravels

approximate location of Mallard Lake Resurgent Dome

Twin Buttes

Twin Buttes ring a steam explosion crater.

Midway Geyser Basin

geysers in cracks of parking area pavement at Biscuit Basin

Old Faithful

valley fill

West Thumb Geyser Basin

Upper Geyser Basin

West Thumb of Yellowstone Lake

West Thumb

20 percent of world's geysers in the Upper Geyser Basin

Shoshone Geyser Basin

Shoshone Lake

rhyolite welded ash flow

N

0   5mi

8km

MADISON — WEST THUMB

Giantess Geyser
National Park Service Photograph.

Shortly before the road crosses Nez Perce Creek north of Fountain Paint Pots, a veneer of glacial sand and gravel caps the Quaternary rhyolite. These gravels were deposited by the most recent Pinedale glacial event sometime around 13,000 years ago. We know these glaciers came after the last volcanic eruption because the gravels sit on top of the lava flow.

Seven and a half miles south of Madison Junction, about half a mile north of the exit of the one-way northbound Firehole Lake drive, the road crosses lake silts made of the hard siliceous remains of single-celled algae called diatoms. Low hills to the north are glacial gravel from both the Bull Lake and the more recent Pinedale glaciers. Twin Buttes, to the southwest, ring a crater from a steam explosion caused by overheated ground water that exploded into steam, blasting out the crater. The explosion may have been caused by a pressure drop when a glacial lake drained suddenly.

The parking area for Fountain Paint Pots is in the middle of the Lower Geyser Basin. A slightly different kind of thermal feature occurs here and also at Mud Volcano (see the Canyon—Fishing Bridge roadguide). The mud pots exist where there is not enough water to support a geyser or a hot spring even though some water is boiling at depth. Shortage of water also permits sulfuric acid to accumulate, and attack the rock, changing it into sticky clay. As steam and gas vapors bubble up through the mud it produces the enchanting bubbling mud pots. See if you can get a picture of one with a blob of mud frozen in mid-air—be prepared to waste a bit of film though, because the bubbles occur sporadically. Be careful not to get burned from a splatter of the hot mud. Here, and in all thermal areas in the park, stay on the board walks.

It is interesting to notice the distribution of paint pots and geysers in this basin. The paint pots are all on high ground around the edge of

113

the basin and must be above the main water table. This higher ground allows for a limited amount of water to make mud pots which are dominated by vapors rather than water. The geysers, on the other hand, are a little lower in the valley below the water table, where there is sufficient water to support the numerous eruptions.

All rocks along the road between the parking area and the entrance to the Firehole Lake drive are of siliceous sinter deposited by various thermal features. The long ridge that makes up the western skyline between the Lower and Midway geyser basins is the 115,000-year-old West Yellowstone rhyolite flow of the Central Plateau member.

The Midway Geyser Basin, 10 miles south of Madison Junction, extends for about a mile along the Firehole River. Even though small, it includes some interesting thermal features. This basin must have an ample ground water supply because it contains Grand Prismatic Spring, the largest hot spring in North America, and a former geyser (Excelsior) that still pumps out a constant flow of over 4000 gallons of boiling water every minute. Contrast this with the spits of water produced by the fumaroles on high ground in the Lower Geyser Basin. Excelsior erupted as a geyser for the last time in 1888. Apparently the last witnessed eruptions were so violent that they tore out all restrictions in the water conduit. This allowed the ground water to flow freely to the surface making a hot spring. The spring will not erupt as a geyser until something again clogs the path to the surface— possibly deposition of siliceous sinter on the walls of the conduit or a closing of the path from movement along a fault during an earthquake. Because both of these events are common occurrences in the geologically active thermal basins, Excelsior should be considered only dormant, and not extinct.

The parking lot at Biscuit Basin shows the persistence of thermal features. It is easy to imagine this whole area as the steaming remains of the giant volcano that is still causing all of the thermal activity and rapid change. Several small steam vents even exit through the asphalt of the parking lot. All bedrock in the area is completely covered by till debris from Pinedale glaciers.

Two miles south of the Biscuit Basin parking lot, is Black Sand Basin with about 12 hot springs and geysers and the cloverleaf entrance to the Upper Geyser Basin and Old Faithful. Geysers are more numerous here than anywhere else in the world because of all the hot cooling rock only a mile or two beneath the surface. In fact, the Upper Geyser Basin alone, within a square mile, includes about 140 geysers or nearly a fifth of the world's geysers.

One drill hole near this area (USGS Lone Star Y-6) found water at a temperature of 358 degrees Fahrenheit only 500 feet below the surface; even hotter water surely exists below drill depth. The high pressure that exists at depth keeps the water from boiling. If some water eventually does seep out to the surface causing a pressure drop, the rest flashes into steam and the expanding steam pushes the entire column of steam and hot water out as a geyser. An additional feature needed to make a geyser is a porous reservoir to store water that can then slowly seep into the main conduit. Glacial sand and gravel fill most of the geyser basins and provide just such a reservoir.

The water stays hot even at the surface. The average water temperature at the Upper Geyser Basin is about 170 to 180 degrees Fahrenheit, hot enough to cause severe burns and even death. And individual features have even hotter water. Water in Old Faithful Geyser can reach 204 degrees Fahrenheit during peak eruption. This is superheated water, hotter than the boiling temperaure at this elevation. Please obey the park signs put up for your own protection. Every year people are badly burned because they wander too close to a geyser or hot spring despite obvious warnings. By wandering off the board walks, you are also likely to shatter brittle sinter deposits, destroying the beauty of the area. Play it safe and watch from a distance.

As the hot water rises to the surface, it dissolves minerals from rocks at depth. These minerals then precipitate out when the water cools on the surface and form picturesque terraces. Where sedimentary rocks underlie the geyser basin, as at Mammoth Hot Springs, the water dissolves calcium from limestone layers and forms travertine at the surface. However, in the central part of the park, no older sedimentary rocks should exist at depth—all were probably melted and destroyed by the volcanic activity. Thus, the water flows only through high silica rhyolite lavas and becomes saturated in silica. Geysers in the central plateau deposit siliceous sinter or geyserite around the vents. Because silica does not dissolve very well, the sinter terraces grow very slowly.

The Upper Geyser Basin is a very good place to watch geysers work and to see many of the features that I have just described. If you are tired of the crowds, cloverleaf, and traditional photos of Old Faithful,

ter deposits
ound Castle
Geyser in the
Upper Geyser
Basin.

Old Faithful Geyser.
National Park Service Photograph.

plan to spend one-half to several days to explore this basin. There are many miles of walks to quiet secluded pools and spectacular geysers where you can really observe the features in peace, away from the crowds, and get some original photographs.

Grotto Geyser, about a mile walk to the northwest of Old Faithful, has a very interesting and complicated series of vents through the sinter. When this geyser first started to form, it may have killed trees growing nearby. These trees became coated with sinter, partially plugging the vent. The mass of sinter and trees now accounts for the complicated grotto of the compound vent. Please don't throw trash into the pools or vents as this may also change the plumbing and damage the features. It is impossible to now enjoy some thermal features that once existed in the park because of the thoughtless actions of previous visitors.

Many of the bright colors around the hot spring pools are bacteria and algae. Bacteria live in the hottest water by the vents, while algae live farther away where the water has cooled a bit. Each species prefers a different temperature of water and this makes the color banding around the pools. Some of the species actually thrive in the scalding water. However, some of the bacteria and algae that live only in the hot water could survive nicely in the cooler water as well, if it were not for fierce competition for food and living space that exists in the cooler water from the many organisms who can tolerate the cooler temperatures. Thus, only a select few species can escape and "beat out" the competition by living in the very hot water. Around the edges of some pools, trees have been killed by the hot water or from drowning. Some of the cells of the trees have been clogged by silica and they have thus started to petrify. By placing small chips of wood in some of the pools, scientists have learned that petrification can occur in just a few months.

The Upper Geyser Basin occurs on the west edge of one of the resurgent domes in the center of the Yellowstone caldera called the Mallard Lake Dome. A new football-shaped area is still swelling today at over an inch per year. If this swelling were to continue long enough, it could presage another volcanic eruption. Changes are ongoing in this basin. The 7.1-magnitude earthquake that occurred in October, 1983, located near Mount Borah, Idaho, 150 miles to the west caused some fascinating changes. Ground movement from this distant tremor must have changed some of the plumbing in the geyser basin because about 37 thermal features around Old Faithful and Geyser Hill experienced significant changes. The one noticed by most visitors was an increase in the average time between eruptions of Old Faithful by about seven minutes. This increased time interval may be because the water does not flow into the reservoir as quickly due to changes caused by the earthquake. However, other geysers that had been dormant for some time have started to erupt again. No one is certain how long these changes will last, but they show the ever-changing geological processes that shape the Yellowstone Country geology.

East of Old Faithful, the road climbs over the Continental Divide. Even though you are driving east you cross to the west side of the divide! For a few miles before the road crosses back to the east side, you are driving in the upper watershed of the Snake and Columbia rivers. From Shoshone Point you can get a good view of some picturesque scenery on a clear day. Pitchstone Plateau, to the south, is the surface of the youngest flow from the caldera. The flow is 70,000-80,000 years old and covers over one hundred square miles. The surface of the flow shows large wrinkles caused as the surface started to gel while the flow was still moving forward. The mountains south of the plateau are the Tetons near Jackson, Wyoming. From here you can also see the hills ringing the Yellowstone caldera that marks the rim and also the Mallard Lake resurgent dome north of Old Faithful.

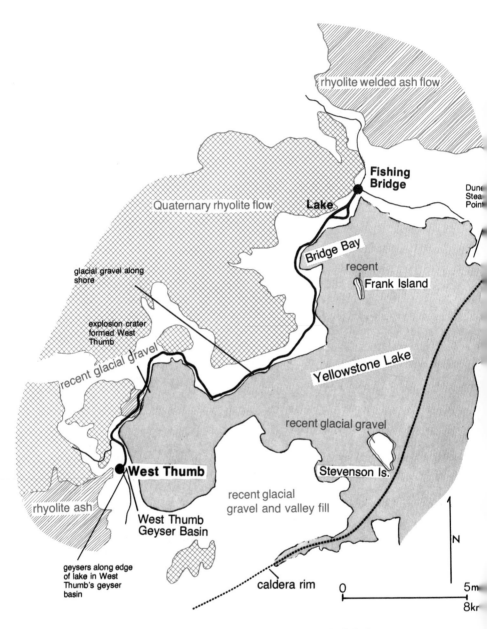

rhyolite welded ash flow

Fishing
Bridge

Dune
Stea
Poin

Quaternary rhyolite flow   **Lake**

Bridge Bay

glacial gravel along
shore

recent

Frank Island

explosion crater
formed West
Thumb

recent glacial gravel

Yellowstone Lake

recent glacial gravel

●**West Thumb**

Stevenson Is.

recent glacial
gravel and valley fill

rhyolite ash

West Thumb
Geyser Basin

geysers along edge
of lake in West
Thumb's geyser
basin

caldera rim

N

0   5m

8kr

WEST THUMB — FISHING BRIDGE

West Thumb—Fishing Bridge
21 mi./34 km.

West Thumb of Yellowstone Lake floods a smaller secondary caldera made by an explosion within the larger Yellowstone caldera after the explosive climax of the third volcanic cycle. Intermittent volcanic activity has occurred since the explosive eruption. One of these separate explosions at West Thumb less than 150,000 years ago left a crater that later filled with water and now makes the embayment on the west of the lake.

Just north of the intersection with the road to the South Entrance, several geysers spout sporadically along the edge of the lake in the West Thumb Geyser Basin. This is one of the smallest of the major geyser basins in Yellowstone and is only a narrow strip for about 1000 feet along the shore of Yellowstone Lake. There are only six geysers in the basin, and most are dormant. The only regular recent activity, despite its name, is from Occasional Geyser that erupts about every half hour. Lone Pine Geyser has shown a bit more activity in the last year or so, but it is not clear if it will continue because it first erupted only in 1974. Fishing Cone is an interesting thermal feature that is a small vent that sticks up out of the water just offshore. Legends tell of fishermen catching fish in the lake and then cooking them in the hot water of the vent while still on the hook! This might have been true in the past, but the water in the cone today is too cool for cooking. Don't try to prove this story by fishing in the area. The sinter crust is thin; you might break through and end up cooking a leg. Fishing is prohibited from the shore here or this very reason. Siliceous sinter around Fishing Cone extends out into the lake to a depth of 5-6 feet. Because geologists think sinter can only form above water, the lake level here must have risen recently, possibly because of tilting associated with ongoing uplift in the center of the Yellowstone caldera.

119

Gravel shores of the lake are a thin veneer of glacial debris left from the ice that once covered all of this area except for the highest peaks. The entire drive from West Thumb to Fishing Bridge is still within the caldera and is along the flat plateau. No good outcrops occur along the road. All rocks are the thick rhyolite lava flows like the one exposed at Firehole Falls. However, here they have been covered by thick pine forests, glacial gravel, and lake deposits. If you walk along the lake shore, it is possible to see various wave-cut benches from different lake stands, and deposits of glacial-fluvial sediments. At Pumice Point, there is volcanic pumice within 20 yards of the road.

If you have a boat, there is a spectacular and complete record of Pinedale and Recent sediments exposed on Dot and Stevenson islands in Yellowstone Lake. The sequence on Dot Island has late Pinedale interglacial lake clays overlain by glacial till. Above these are glacial-fluvial sands, lake varves (thin yearly layers), and ash from an eruption of Glacier Peak in the North Cascades of Washington state. At the very surface are beach sands and windblown sand dunes. One area to see the sand dunes is along the lake near Steamboat Point.

Apparently there are few faults and cracks in the rocks here. No geyser basins exist between the West Thumb Basin and Fishing Bridge. Fractures would allow any hot water to rise to the surface forming a geyser or hot spring. Also, hot water discharge may be beneath Yellowstone Lake.

The Yellowstone River drains from Yellowstone Lake at Fishing Bridge. The Grand Canyon of the Yellowstone (see the Roosevelt—Canyon roadguide) occurs where the river cuts down at the edge of the high plateau in rock weakened by all of the hot water.

Wave-cut terraces above lake level show that either there was once more water in the lake during glacial meltback or recent uplift has occurred. Some of the terraces are no longer level, but slope gently. Because water cuts flat-lying shorelines, these have been tilted. By measuring the amount of tilt and by dating the terrace geologists can determine that uplift of nearly an inch per year is occurring along a football-shaped dome to the north.

West Thumb—South Entrance
22 mi./35 km.

The road from West Thumb to the South Entrance starts out in the Yellowstone caldera now filled with rhyolite flows and then crosses onto welded-ash units that cover older Mesozoic and Paleozoic sedimentary rocks. Few good outcrops are visible along the first part of the road because the flat-lying lava flows are covered by thick lodgepole pine forests.

South of Grant Village, the road crosses to the west side of the Continental Divide and into the Snake River watershed. The plateau is so flat you would probably not realize that you had crossed the pass if it were not for the sign. At the south end of Lewis Lake, the road crosses the edge of the plateau that makes up the rim of the Yellowstone caldera. As the Lewis River drains the lake and flows over the rim, it has eroded the soft rocks leaving the hard ones to make a small waterfall. The 20- to 50-foot high Lewis Falls provides an excellent exposure of one of the rhyolite flows with a breccia zone near the top. Pitchstone Plateau, out of sight to the west, is made of one giant flow that covers over 100 square miles. The plateau is of interest because it is covered by the youngest volcanic flow from the caldera. The flow was produced 70 to 80 thousand years ago and still shows pressure ridges caused by wrinkling of cooling lava at the surface of the flow while it was still moving.

A few miles to the east at the northwest edge of Heart Lake along Witch Creek, is the Heart Lake Geyser Basin. This basin can be reached only by an eight-mile hike and is probably the most remote and least visited geyser basin in the park. Bryan (1979) describes this basin and each of the thermal features.

South of the Lewis River Falls, the road descends from the rim onto welded-ash flows that roared out of the caldera. The Lewis River Canyon cuts through and exposes some of these flows in the canyon walls. Here the ash flows overlie older rhyolite lava flows that, in

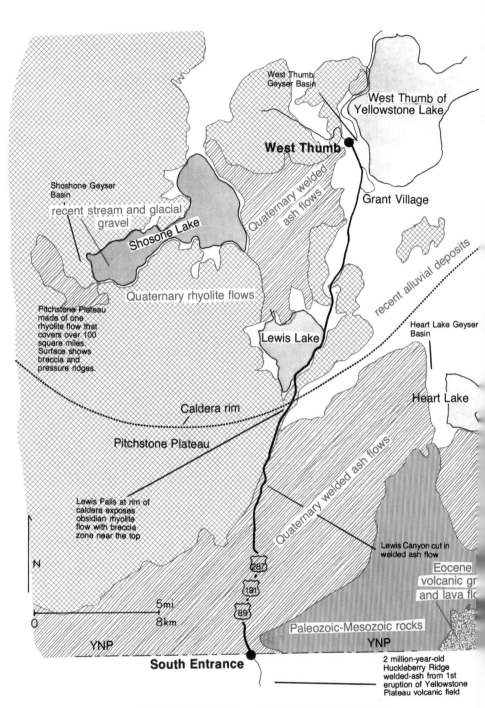

West Thumb Geyser Basin

West Thumb of Yellowstone Lake

West Thumb

Grant Village

Shoshone Geyser Basin

recent stream and glacial gravel

Shosone Lake

Quaternary welded ash flows

recent alluvial deposits

Quaternary rhyolite flows

Pitchstone Plateau made of one rhyolite flow that covers over 100 square miles. Surface shows breccia and pressure ridges.

Lewis Lake

Heart Lake Geyser Basin

Heart Lake

Caldera rim

Pitchstone Plateau

Lewis Falls at rim of caldera exposes obsidian rhyolite flow with breccia zone near the top

Quaternary welded ash flows

Lewis Canyon cut in welded ash flow

Eocene volcanic gr and lava fl

N

287

191

89

Paleozoic-Mesozoic rocks

5mi

8km

0

YNP

YNP

South Entrance

2 million-year-old Huckleberry Ridge welded-ash from 1st eruption of Yellowstone Plateau volcanic field

WEST THUMB — SOUTH ENTRANCE

turn, lie on Mesozoic and Paleozoic sedimentay rocks. None of these older rocks is exposed along the road but some can be reached by a hike up into the Red Mountains area. Just south of the park, is the southern remnant of the caldera rim from the caldera-forming explosion of the first eruptive cycle on the Yellowstone Plateau volcanic field, 2 million years ago. Huckleberry Mountain (elev. 9,615 feet), the large hill southeast of the entrance, is made of the Huckleberry Ridge Tuff erupted during this explosion.

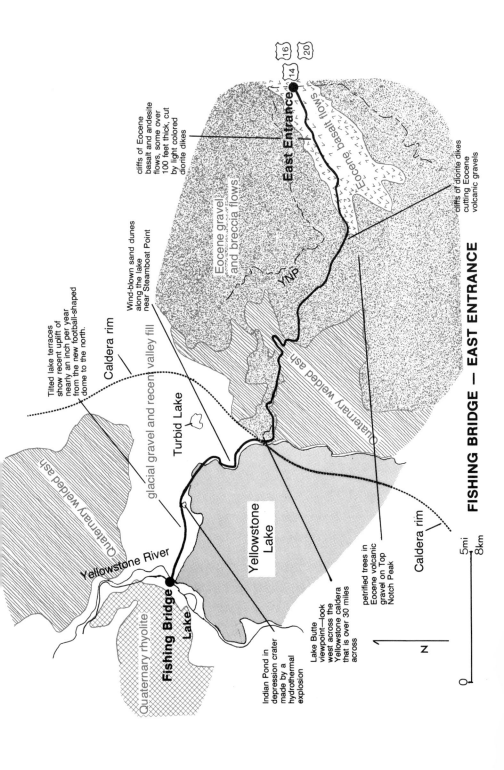

FISHING BRIDGE — EAST ENTRANCE

cliffs of Eocene basalt and andesite flows, some over 100 feet thick, cut by light colored diorite dikes

East Entrance

14 16 20

Eocene basalt flows

cliffs of diorite dikes cutting Eocene volcanic gravels

Eocene gravel and breccia flows

YNP

Quaternary welded ash

Tilted lake terraces show recent uplift of nearly an inch per year from the new football-shaped dome to the north.

Caldera rim

Wind-blown sand dunes along the lake near Steamboat Point

glacial gravel and recent valley fill

Turbid Lake

Quaternary welded ash

Yellowstone River

Fishing Bridge

Quaternary rhyolite

Yellowstone Lake

Indian Pond in depression crater made by a hydrothermal explosion

Lake Butte viewpoint—look west across the Yellowstone caldera that is over 30 miles across

petrified trees in Eocene volcanic gravel on Top Notch Peak

Caldera rim

N

0   5mi
     8km

Fishing Bridge—East Entrance
27 mi./43 km.

Fishing Bridge and Yellowstone Lake lie on the east edge of the Yellowstone caldera. The lake formed as water filled part of the depression left by the last explosion 600,000 years ago and the low seam between the Yellowstone Plateau and the Eocene Absaroka Volcanics. Gravel around the edge of the lake is glacial till and outwash showing that the area was once covered by a glacier after the last eruption of the caldera.

Indian Pond (formerly Squaw Lake) and Turbid Lake east of Fishing Bridge are two of at least ten hydrothermal explosion craters in Yellowstone Park. The explosions were caused when hot ground water above the boiling point expanded after a rapid decrease in confining pressure. High pressure from water in a glacial lake may have kept the superheated ground water from boiling—even up to 450 degrees Fahrenheit or hotter. Glacial lakes have a tendency to drain quickly because ice does not make a very stable long-lasting dam. The resulting pressure drop from the sudden draining would have allowed the water to flash into steam, blasting out a sizable crater.

Rocks between Fishing Bridge and the East Entrance of Yellowstone National Park record two different periods of volcanism. Around Fishing Bridge, glacial gravel covers rhyolite ash and welded-ash flows from the Quaternary eruptions of the caldera. These recent volcanic rocks are rich in silica and are rhyolites. The older period of volcanism is recorded by the 50 million-year-old andesitic Absaroka Volcanics.

Lake Butte Viewpoint is on the rim of the giant volcano from the third cycle of eruptions of 600,000 years ago that produced much of the rhyolite flows throughout the Yellowstone Country. The short side trip is well worth the drive to look west across the 30-mile wide caldera. The rim lies beyond the skyline on the far western side of the

Indian Pond, a crater formed by a large steam explosion.

lake. The caldera is not very deep because rhyolite lava oozed out to fill the depression left by the main explosion. The Red Mountains, and Mount Sheridan, visible to the south, are uplifted Paleozoic rocks that mark the edge of the caldera. Roadcuts along the road to the viewpoint expose 50 million-year-old Absaroka conglomerate and an intrusion of Eocene diorite.

From the viewpoint to Sylvan Pass, the Quaternary volcanic ash flows lap onto older 38-50 million-year-old deposits from the Absaroka volcanoes. These Eocene volcanic deposits are of andesite, a type of lava intermediate between basalt and rhyolite. The lava of the Eocene volcanic rock here contains much iron and a dark magnesium-rich pyroxene mineral. This black mineral makes the lava flows appear very dark brown to purple.

Several different kinds of Eocene volcanic rock occur along the road in the vicinity of Sylvan Pass. The pass must have been near one of the Eocene volcanic centers because much of the rock is andesitic lava flows and a very angular breccia made up of fragments with very sharp edges that could not have moved very far. Away from the pass, both to the east and to the west, the flows and breccias grade into mudflows and stream gravels that look like the rocks burying the fossil forests in the Lamar River valley. These rocks were deposited by the same processes, but were erupted from different vents. Vents here in the southern part of the park erupted a few million years after ones around Cooke City and Mount Washburn.

Some of the water-sedimented volcanic debris also buries petrified trees around Sylvan Pass. Good examples are on Top Notch Peak south of the park road.

Near the pass, a large cliff on the south side of the road is a dike of light-colored diorite that was injected into the andesitic layers. This must have happened as one of the volcanoes erupted. In fact, this dike may have been the plumbing that supplied magma to the volcano.

Just east of the entrance, cliffs on the north side of the road are made of dark lava flows interbedded with breccia. These were later cut by more dikes of diorite.

126

East Entrance—Cody
53 mi./85 km.

From the East Entrance of Yellowstone National Park 30 miles or so east to just west of Wapiti, the road crosses Eocene volcanic sands and gravels of the Absaroka Volcanic Supergroup. These rocks were deposited in the same environments as those that bury the fossil forests in the northern part of the park (Sepulcher and Lamar River formations) and belong to the same package of volcanic/sedimentary rock, the Absaroka Volcanic Supergroup. However, volcanic rocks here were deposited from debris erupted from vents farther south and several million years later than eruptions to the north. Some petrified trees exist in volcanic gravel along the road but not as many as in formations exposed in the northern part of the park.

From Pahaska Tepee east, the valley of the Shoshone River has a narrow cross section. No major glacier flowed down this valley to gouge out a rounded profile. However, gravel washed downstream by the river during high discharge caused by the melting of glacial ice has formed the terraces beneath the town of Cody. Some of the side valleys, though, contained glaciers, and are blocked by terminal moraines.

On the north side of the road, near the mouth of Blackwater Creek, is a small cave carved by the river as it eroded through soft volcanic gravel and sand. Carbon-14 dating of bits of wood buried in the river valley show that the river has been at its present level for the past 10,000 years, since the cave was formed. The dry cave served as a temporary home for early man. The cave is called Mummy Cave because the mummified body of a man who died 1300 years ago, around 680 A.D., was found in it. Archaeologists have identified about 38 separate occupations of the cave covering a time span of nearly 9000 years, from 7280 B.C. to 1580 A.D. Cliffs above the cave provide a good place to see Eocene volcanic gravels which contain a few scattered petrified stumps and logs.

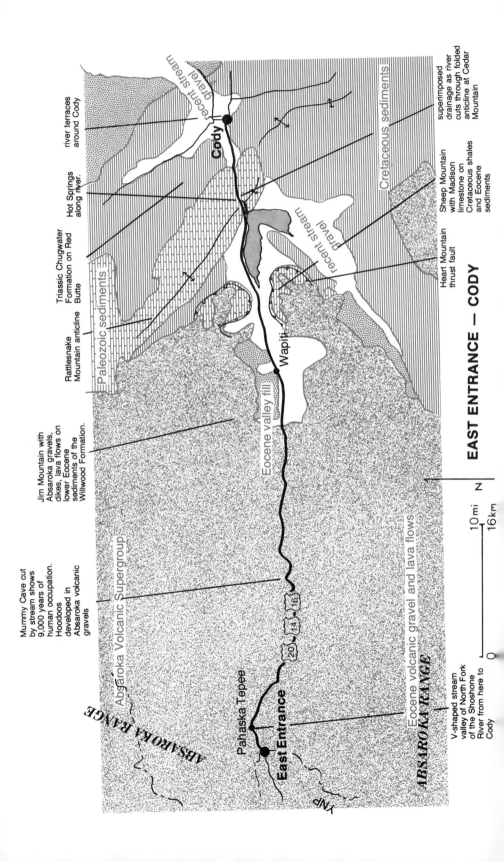

EAST ENTRANCE — CODY

Mummy Cave cut by stream shows 9,000 years of human occupation. Hoodoos developed in Absaroka volcanic gravels

Jim Mountain with Absaroka gravels, dikes, lava flows on lower Eocene sediments of the Willwood Formation.

Rattlesnake Mountain anticline

Triassic Chugwater Formation on Red Butte

Hot Springs along river.

river terraces around Cody

recent stream gravel

Cody

Paleozoic sediments

Cretaceous sediments

superimposed drainage as river cuts through folded anticline at Cedar Mountain

Sheep Mountain with Madison limestone on Cretaceous shales and Eocene sediments

Heart Mountain thrust fault

Wapiti

Eocene valley fill

recent stream gravel

Absaroka Volcanic Supergroup

ABSAROKA RANGE

Eocene volcanic gravel and lava flows

ABSAROKA RANGE

Pahaska Tepee

20 · 14 · 16

East Entrance

V-shaped stream valley of North Fork of the Shoshone River from here to Cody

YNP

N

10 mi
16 km

0

From Mummy Cave to Wapiti, striking, tall, thin pinnacles have weathered in the easily eroded conglomerate. Weathering occurs along vertical cracks in the horizontal beds of conglomerate and leaves the tall spires. In places, huge washtub-size cobbles sit atop columns making them into hoodoos. The large cobble serves as a cap that protects the shaft from splashing raindrops. Good examples can be seen all along the road—one especially good place to see them is at a cliff called the Holy City.

Where the Shoshone River valley starts to widen west of Wapiti, light-colored horizontal beds of water-deposited rhyolite ash appear at river level, the early Eocene Willwood Formation. The river is wider here because the soft ash erodes more easily than the younger conglomerates, thus allowing the river to develop a wide flood plain. From here to Cody the river erodes into successively older sedimentary rocks down to the Precambrian basement.

At Wapiti, the river has eroded through the Willwood Formation and into underlying Cretaceous shales at river level. The prominent Chinese Wall on the west edge of Wapiti is a long vertical dike, magma that hardened after it was squeezed into a vertical crack in the horizontal sediments now removed by erosion. Jim Mountain, to the north, contains Absaroka volcanic sediments on the cliff faces resting on the light-colored Willwood Formation at the base. Notice how the Willwood Formation erodes into low, smooth-looking hills, while the resistant gravel holds up an almost vertical cliff face. Clasts of the Absaroka volcanic gravels are made of dark andesite. In contrast the ash grains of the Willwood Formation are made of pale rhyolite. The Absaroka cliffs are even darker because they contain many black basalt flows and dikes.

West of Wapiti, several of the hills north and south of the road contain blocks of the Heart Mountain thrust fault. One of these, Sheep Mountain, south of the road, is a block of 300 million-year-old Madison Limestone resting on 55 million-year-old sediments of the Willwood Formation. This is the same fault that can be traced from the Cooke City area to Cody. The transported blocks must have slid out over the Eocene landscape from somewhere in the vicinity of Cooke City.

At the west edge of Buffalo Bill Reservoir, a large block of the Precambrian basement was tilted and thrust west to the surface with a gentle slope back to the east. The once flat-lying Paleozoic and Mesozoic sediments now drape over the uplift forming the Rattlesnake Mountain anticline. The sediments dip very steeply to the west on the west side of the mountain and gently to the east on the east side.

You can tell something about when the uplift occurred by studying the course of the Shoshone River. As you drive by the reservoir, notice

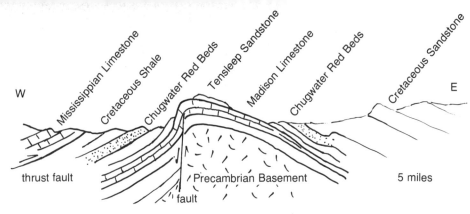

*Diagramatic cross section through the Rattlesnake Mountain
anticline along US 14/16/20 (Redrafted and modified from the profile
of Field, 1932).*

that the river flows through a narrow canyon. This canyon was cut by
the river dividing Cedar Mountain to the south from Rattlesnake
Mountain to the north. It at first seems difficult to understand why
the river would have cut through a whole mountain when, by eroding
only a few soft shales, it could easily have flowed around the south
end of Cedar Mountain. The solution is that the river is older than the
mountain! If the mountain had existed before the river, the river
would surely have flowed to the south. However, the river established
a valley while the sediments were still flat-lying. Then, the river
became trapped in its own valley and eroded the canyon as the
mountain rose.

*The Rattlesnake Mountain anticline as seen from Cody. Notice how the
Shoshone River has divided the anticline into Cedar Mountain (left)
and Rattlesnake Mountain (right).*

Along the west edge of the uplift, the resistant sedimentary layers
that were tilted to nearly vertical have been eroded into wedge-
shaped triangles called flatirons, because they look like the point of

130

Banded gneiss in Precambrian basement rocks in the core of the Rattlesnake Mountain anticline.

an old-fashioned flatiron. The first beds raised up at the edge of the canyon are bright red beds of the Triassic Chugwater Formation. This red formation is a key geologic marker throughout Wyoming because it is very easy to identify. Below the Chugwater Formation are various formations deposited during Paleozoic time. The thick white Madison Limestone makes the most prominent flatirons. The first two tunnels from the west cut through the Madison Limestone. To identify the rest of the formations consult the generalized stratigraphic column on page vi.

At the dam, the river has cut into the Precambrian basement block. Like all Precambrian basement rocks in the Yellowstone Country, this block is made out of pink granites, banded gneiss, and dark amphibolite schist. The road follows the Precambrian basement for about half a mile. At the east end of the last tunnel there is a good place to stop and look at the contact between the 2.7 billion-year-old Precambrian basement and the 550 million-year-old Flathead Sandstone, and to look back upstream at a well-exposed wall of banded gneiss. The Flathead Sandstone has pebbles of pink feldspar near the contact, showing that it is made of grains eroded from the basement, and becomes finer-grained and crossbedded higher up. Driving east you will now encounter, in reverse order, all of the Paleozoic formations that you passed through on the west limb of the anticline. See if

Cross-bedded Flathead Sandstone resting on Precambrian granite at the east end of the easternmost tunnel in the Rattlesnake Mountain anticline.

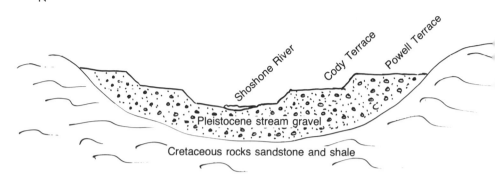

Cross section through the Cody area showing old stream terraces along the North Fork of the Shoshone River. Older terraces are high above the present river; younger ones lower in the valley.

you can tell when you have driven through all of the Paleozoic formations by recognizing the Triassic Chugwater Formation made of bright red sandstone and shale layers. Notice that the beds on the east side of the mountain dip to only 10 degrees or so into the Bighorn Basin to the east.

There are four interesting hot springs along the river between the east edge of the reservoir and Cody. The first white person to report these springs was John Colter in 1807 or 1808. John called the river the "Stinkingwater"—presumably from the sulphur odor emitted by the springs. His description of the pools made his fellow fur trappers refer to the area as Colter's Hell—a name often wrongly attributed to geyser basins in Yellowstone National Park.

Cody is on the Pleistocene floodplain gravels of the Shoshone River. Notice the matching terraces on either side of the river. The upper terrace is the oldest and was deposited when the river was at that elevation. As the river eroded the valley, it formed a new terrace at a

Well-rounded stream gravel in the Powell Terrace above the main part of Cody. This gravel was deposited by outwash from melting Bull Lake glaciers.

lower level. Because the river is still cutting, a recent terrace exists near its present level. The oldest terrace is about 60 feet above the main part of town and is called the Powell Terrace. The Powell Terrace correlates with the Bull Lake glacial event and represents gravel carried here during the glaciation. Downtown Cody is built on the younger Cody Terrace from Pinedale glacial outwash. Many roadcuts around town allow you to examine the river gravel that is made of walnut- to baseball-size well-sorted cobbles rounded by stream flow.

VII
Glossary

Alluvial fan: A fan-shaped deposit of sediment produced where a stream flows from a mountain out onto a valley floor. Deposited by both mudflows and streams.

Amphibolite: A very dark-colored metamorphic rock that has a lot of minerals rich in iron and magnesium.

Andesite: A common volcanic rock with a silica content intermediate between high-silica rhyolite and low silica-basalt. Occurs as flows, breccia, ash or a water-deposited debris. May be various shades of dark brown, green, or purple.

Anticline: Layers of rock folded into an arch. Oldest rocks occur in the center.

Arête: A knife-like ridge formed by the erosion of two series of glaciers, one on either side of the ridge.

Basalt: A typically black fine-grained volcanic rock low in silica and rich in iron bearing minerals. This rock is not viscous when molten and can flow for a long distance in thin flows. Often contains post-pile looking columnar joints caused by contraction while cooling.

Basement: Igneous and metamorphic rocks, Precambrian in age in the Yellowstone Country, that underlie the sedimentary rocks.

Breccia: A rock made of fragments with sharp edges. Often forms at the top of lava flows. As the top of the flow hardens it breaks and shatters as the flow continues to move.

Bull Lake glaciation: The name given to the glacial advance that occurred between 160 and 130 thousand years ago.

Caldera: A very large volcanic crater generally formed by collapse. Collapse is caused by the magma chamber rapidly expelling most of its magma. The roof of the chamber then collapses leaving a large caldera. The Yellowstone caldera is over 35 miles across.

Cenozoic: The most recent geologic era from 65 million years ago to today. For Period and Epoch subdivisions, consult the time scale on page vi.

Cirque: A very steep-walled amphitheater eroded near the top of a mountain where a glacier formed and started to flow plucking away some of the bedrock. Several cirques may coalesce to form an arête or a horn.

Clast: A particle of sediment, of any size, that has been transported by running water, wind, gravity fall, glaciers, volcanic ejections, etc.

Conglomerate: A sedimentary rock made of gravel-size clasts that range from pea- to car-size.

Country Rock: The rock of any composition around an intrusion. Country rock is always older than the intrusive rock.

Cretaceous: The period of geologic time between 140 and 65 million years ago. A subdivision of the Mesozoic Era.

Diabase: A dark-colored low-silica intrusive rock, chemically like basalt, that often forms dikes and sills.

Dike: A thin intrusion of igneous rock of any composition that is nearly vertical or cuts across layers of the sedimentary rock at an angle.

Diorite: A coarse-grained intrusive igneous rock that cooled deep below the surface of the earth. The rock is very dark-colored because it contains almost no quartz.

Drumlin: A streamlined hill of glacial till deposited and shaped by moving glacial ice. Usually teardrop-shaped with the blunt end facing "upstream."

Eocene: The Epoch of geologic time between 55 and 35 million years ago. A subdivision of the Tertiary Period and Cenozoic Era. See the stratigraphic column on page vi.

Erratic boulder: A boulder carried by a glacier and then dumped out in the middle of fine-grained sediments. Often the boulders are of a very different rock type than what exists close to where they occur showing that they must have been moved a long distance.

Fault: The plane along which two layers of rock have moved. A fracture or crack where the rocks on each side have moved relative to each other.

Feldspar: A very common family of rock-forming minerals. Sodium-calcium feldspar, or plagioclase, is chalky or milky white. Potassium feldspar is dull chalky pink. Both can be identified readily in any of the Precambrian granites in the Yellowstone Country.

Flatiron: Triangular-shaped rock layers torn and tilted on edge by the uplift of a basement block.

Formation: The basic unit for naming rocks that are distinct from the surrounding rocks and large enough to be mapped.

Fumarole: A thermal feature that forms where there is a boiling water table but too little water to make a hot spring or geyser. Instead, a vent forms that produces mostly steam.

Gabbro: A coarse-grained dark-colored rock low in silica. The chemical equivalent of basalt. It differs in that it cooled slowly at depth rather than quickly at the surface.

Geyser: A column of steam or water that periodically shoots out of the ground under pressure from the heating of the water at depth.

Geyserite: Siliceous sinter. Layers of silica deposited around the vent of a geyser or hot spring.

Gneiss: A metamorphic rock formed by the squeezing and flowage of older rocks. Gneisses are always banded—often with quartz-rich layers alternating with dark schist layers.

Graded bed: A layer of sedimentary rock deposited by moving wind or water with the largest particles at the bottom and smaller ones at the top.

Granite: A coarse-grained igneous rock that cooled deep below the surface. High in silica and quartz, the chemical equivalent of rhyolite. It differs in that it cooled slowly at depth rather than quickly near the surface. Contains about 30% quartz, 60-70% feldspar, and a few dark mica, hornblende, and other mineral grains.

Hogback ridge: A long linear ridge formed by the tilting of a resistant layered sedimentary rock up on edge.

Horn: A steep spired mountain. The steep spire was carved by glaciers eroding cirques on all sides. The most famous horn is the Matterhorn in Switzerland. Pilot, Index, and Electric peaks are a few of the many horns in the Yellowstone Country.

Hot spring: A pool of hot water warmer than the average annual air temperature. The water is heated when ground water soaks deep into the ground and comes in contact with hot rocks. Where the water flows freely to the surface, a hot spring develops. If the path is obstructed and the water superheated above the boiling temperature, a geyser may occur instead. Some geologists make a further distinction between a warm spring that is below body temperature and restrict the use of hot spring for a pool of water warmer than body temperature.

Igneous rocks: Rocks that form from the cooling and hardening of molten magma. Plutonic igneous rocks are coarse-grained because they cooled slowly at depth. Volcanic igneous rocks are fine-grained and cooled quickly on the earth's surface.

Intrusion: Igneous rock that was injected as magma into older rocks. Large bodies make batholiths and stocks. Thin stringy layers are dikes or sills.

Joint: A crack in a rock where no movement has occurred. If the rocks on either side had moved it would be a fault.

Kettle lake: A small lake that forms in a depression made as the ground collapsed after a buried block of glacial ice melted. The blocks of ice are left behind as glaciers retreat and the blocks become buried by glacial outwash sediment.

Laminations: Very thin, usually less than an inch or so, layers of sedimentary rock.

Laramide: The name given to the mountain-building episode that created the Rocky Mountains during the late Cretaceous and early Tertiary periods around 50 million years ago.

Lava: Magma that flows out of a volcano onto the surface of the earth and loses its dissolved gaseous content.

Lithophysae: Concentric subcircular cavities in glassy rhyolite and obsidian flows. Thought to have been gas bubbles in the lava that have since been lined by mineral deposits. Occur abundantly in the rocks at the base of Obsidian Cliffs in Yellowstone National Park.

Load casts: Structures formed when heavy sand is deposited on and sinks into soft soupy silt or clay.

Magma: Liquid or molten rock. Rocks high in silica, like granite and rhyolite, melt at around 1300-1650 degrees Fahrenheit (700-900 degrees Celsius) and rocks low in silica, like basalt, melt at around 2200 degrees Fahrenheit (1200 degrees Celsius). Magma is called lava when it flows out onto the earth's surface.

Mesozoic: The era of geologic time between 230 and 65 million years ago. Subdivided into three periods, the Triassic, Jurassic, and Cretaceous. See the time scale on page vi.

Metamorphic rocks: Rocks that have been changed from other rocks by heat and pressure, but without melting.

Mica: A common mineral that, because of its molecular arrangement, splits apart into thin leaves or pages. Muscovite is white or clear and biotite is black mica.

Moraine: Long ridges of glacial sediment called till deposited along the sides and toes of glaciers.

Nick point: A steep cliff that forms where a stream cannot erode a hard resistant layer as fast as a softer one. A waterfall results.

Obsidian: Shiny, dense, black volcanic glass. The rock is glassy because it has no crystals, either because it cooled very quickly or because the magma was very dry.

Orogeny: A period of intense folding, uplift, and mountain building.

Paleozoic: The era of geologic time between 570 and 230 million years ago. The earliest Paleozoic rocks contain the first abundant hard-shelled fossils. See page vi for subdivisions into seven periods. In order from old to young, they are: Cambrian, Ordovician, Silurian, Devonian, Mississippian, Pennsylvanian, and Permian.

Pinedale: The most recent glacial advance that occurred between 70 and 13 thousand years ago. The most prominent glacial features in the Yellowstone Country are from Pinedale glaciers.

Plutonic: Igneous rock that cooled at a great depth below the earth's surface.

137

Precambrian: The earliest portion of earth's history from the formation of the earth about 4.5 billion years ago to the start of the Paleozoic Era about 570 million years ago. Rocks deposited during this time have few fossils, mainly algae, bacteria, worms. In the Yellowstone Country they are represented mainly by unfossiliferous granites and gneisses.

Pumice: A lightweight volcanic rock with so many gas cavities (vesicles) that it will float on water.

Quartz: A clear glassy mineral made of silicon and oxygen. A very common mineral.

Quaternary: The most recent period of geologic time from 2 million years ago to today. A subdivision of the Cenozoic Era—see the time scale on page vi.

Resurgent dome: An area that was uplifted in the form of a dome by the intrusion of underlying magma soon after formation of a caldera. There are two extinct resurgent domes in the Yellowstone caldera. In addition, there is another that is now swelling at over half an inch per year. If the swelling continues long enough, it might be the precurser to a volcanic eruption.

Rhyolite: A volcanic rock formed from the cooling of lava rich in silica. The chemical equivalent of granite. Even when molten, rhyolite lava is very viscous and stiff. Sometimes flows as a lava flow. Commonly erupts explosively and flows as a dense cloud of ash that, when cool, forms a welded-ash flow tuff.

Sedimentary rocks: Rocks that result from the consolidation of loose particles of lime, clay, silt, mud, sand, or gravel eroded and transported from pre-existing rocks and deposited in streams, lakes, deltas, beaches, and oceans. Common types include conglomerate, sandstone, shale, and limestone.

Scarp: A steep cliff left at the back of a landslide or from movement along a fault.

Schist: A metamorphic rock formed from heating of other rocks under high pressure caused by deep burial. These conditions allow new minerals, such as mica, to grow. Normally schists split apart along the planes of these new minerals.

Scoria: Moderately dense lava with numerous cavities from expanding gas bubbles. Denser than pumice.

Scree: Loose rubble of angular rock debris at the base of a cliff or steep slope.

Shale: A kind of sedimentary rock made of consolidated silt and clay that splits apart along the original bedding planes.

Silica: The second most common constituent of the earth's crust. Silicon combined with oxygen (the most abundant element) and its pure form makes quartz. Rocks with abundant silica are said to be siliceous.

Sill: An intrusive rock of any composition that was squeezed into layered rock parallel to the layers.

Sinter: A siliceous mineral deposited around the throat of a geyser or hot spring. The silica was dissolved from silica-rich rocks by the hot water and precipitated out of solution as the water cooled.

Stock: A small body of plutonic igneous rock. May be a small body that cooled at depth or is often magma that cooled in the throat of, or at least near, a volcano.

Superimposed drainage: A river that cuts deep into underlying bedrock and is trapped in its valley. Often is a stream that cuts through a mountain when it now appears that it would have been easier for the stream to just flow around. However the stream is older than the uplift and was trapped in its bed as the uplift slowly occurred to form the mountains.

Syncline: Layered rocks folded into a valley or bowl with the layers concave upward. Youngest layers are in the center.

Thrust fault: A low angle fault that can shove old rocks up over younger ones.

Till: Non-sorted glacial debris composed of all grain sizes from clay to boulders. Till is the rock type that makes moraine landforms.

Travertine: Limestone hot spring deposits. Often the calcium carbonate is precipitated from the cooling of hot ground water rich in calcium carbonate by algae and bacteria.

Talus: Large loose angular blocks of rock debris at the base of a cliff or steep slope.

Tufa: A calcareous rock deposited around hot springs. More or less synonymous with travertine.

Tuff: Hardened volcanic ash formed by the consolidation of small pieces of shattered magma or volcanic rock.

Unconformity: An erosional surface between older and younger rocks. Indicates missing time. An angular unconformity can result when layers beneath the unconformity were tilted to an angle before erosion and deposition of younger horizontal layers.

Vesicle: A gas bubble cavity in a lava flow.

Volcaniclastic: A sedimentary rock with particles (clasts) of volcanic composition derived from the erosion of volcanic rocks.

Volcanic rocks: Igneous rocks that cooled from magma at the surface of the earth.

Welded-ash flow tuff: A dense layer of rock that formed as hot ash particles welded themselves together. Often of rhyolitic composition.

Yellowstone Plateau volcanic field: A large area of volcanic activity over the last 2.5 million years in the Yellowstone Country. Three caldera-forming explosive climaxes occurred at 2, 1.3, and .6 million years ago. These events have helped to shape the Yellowstone plateau.

VIII
Outside Reading

Because this book is only a brief introduction to the fascinating and sometimes extremely complicated geology of the Yellowstone Country, I have included a list of books and papers for those who desire to know more of the local geology. This is an annotated list of fairly general treatments of various aspects of the area and is by no means complete. Many of the listed books can be purchased in bookstores in the Yellowstone Country and most are readily available at good libraries. Yellowstone National Park maintains a library that is open to the public in the Albright Visitor Center at Mammoth for those who wish to read some of these works while in the Yellowstone Country. The Mammoth library has most of the references listed here. If you want to dig even further into the primary scientific literature, each of these references cites specific papers and researchers involved.

American Association of Petroleum Geologists, 1972, *Geological Highway Map of the Northern Rocky Mountain Region:* AAPG Tulsa, Oklahoma. This is a good general map of the entire Yellowstone Country. It also contains a lot of easy-to-understand text, cross sections, and a summary of the tectonic history of the area.

Bargar, K. E., 1978, *Geology and thermal history of Mammoth Hot Springs, Yellowstone National Park, Wyoming:* U.S. Geological Survey, Bulletin 1444. This paper provides an excellent easy-to-understand discussion of the more than 100 hot springs around Mammoth.

Blackstone, D. L., Jr., 1971, *Traveler's Guide to the Geology of Wyoming:* Wyoming Geological Survey, Bulletin 55. Provides a good summary of the geology of Wyoming, including the Yellowstone Country.

Bryan, T. S., 1979, *The Geysers of Yellowstone:* Colorado Associated University Press, 225 pages. This book is a must for any "geyser gazer." It contains not only an excellent discussion of the geyser basins and the geologic reasons for geysers in Yellowstone, but a description of just about all of the 320 geysers in the park.

Chafetz, H. S. and Folk, R. L., 1984, "Travertines: depositional morphology and the bacterially constructed constituents": *Journal of Sedimentary Petrology*, v. 54, p. 289-316. This research paper discusses the role of bacteria in depositing travertine limestone around hot springs.

Christiansen, R. L., 1984, "Yellowstone magmatic evolution: its bearing on understanding large-volume explosive volcanism": *Explosive Volcanism: Inception, Evolution, and Hazards*, p. 84-95, National Academy Press, Washington, D.C., 176 pages. In this paper the author describes the history of Quaternary explosive volcanism in the Yellowstone Country.

Christiansen, R. L. and Blank, H. R., 1972, *Volcanic stratigraphy of the Quaternary rhyolite plateau in Yellowstone National Park:* U.S. Geological Survey, Professional Paper 729B. This paper describes the three eruptions and the associated lavas from the central Yellowstone caldera.

Dorf, E., 1964, "The petrified forests of Yellowstone Park": *Scientific American*, v. 210, p. 106-14.

Dorf, E., 1980, *Petrified forests of Yellowstone:* U.S. Government Printing Office, Washington, D.C., 12 pages. In these two publications, the late Dr. Erling Dorf presents a slightly different interpretation of the fossil forests than the one I have developed in my own research. However, these papers make excellent reading for those interested in the fossil forests and provide one of the better summaries of the fossil plants.

Fisk, L. H., 1976, *The Gallatin "petrified forest":* Montana Bureau of Mines and Geology, Special Publication 73, p. 53-72. This paper contains one of the few, and the most recent of, published works on the petrified forests in the Sepulcher Formation in northwest Yellowstone and southwestern Montana.

Fischer, W. A., 1960 (reprint 1970), *Earthquake!:* Yellowstone Library and Museum Association. A discussion of the many earthquakes in the Yellowstone Country, including the large 1959 Hebgen Lake quake that altered many of the thermal features in the park.

Fritz, W. J., 1980, "Reinterpretation of the depositional environment of the Yellowstone fossil forests": *Geology*, v. 8, p. 309-13.

Fritz, W. J., 1980, "Stumps transported and deposited upright by

Mount St. Helens mudflows": *Geology*, v. 8, p. 586-88.

Fritz, W. J., 1982, *Geology of the Lamar River Formation, northeast Yellowstone National Park:* Wyoming Geological Association Guidebook, p. 73-101. In these papers I present my research on the fossil forests and compare them to modern events around Mount St. Helens.

Harris, A. G. and Tuttle, E., 1983, *Geology of National Parks:* (Third edition), Kendall/Hunt Publishing Company, Dubuque, 554 pages. This book contains a good summary of the geology of all of the national parks, including Yellowstone National Park, in non-technical language.

Harris, D. V., 1980, *The geologic story of the National Parks and Monuments:* (Third edition), John Wiley and Sons, New York, 322 pages. A good general description of the geology of all of the national parks and monuments, including Yellowstone National Park.

Hildreth, W., Christiansen, R. L. and O'Neil, J. R., 1984, "Catastrophic isotopic modification of rhyolitic magma at times of caldera subsidence, Yellowstone Plateau volcanic field": *Journal of Geophysical Research*, v. 89, no. B10, p. 8339-69. A thorough technical discussion of events that formed the Yellowstone Plateau volcanic field.

Keefer, W. R., 1971, *The geologic story of Yellowstone National Park:* U.S. Geological Survey, Bulletin 1347. An excellent non-technical summary of the geology of the Yellowstone National Park.

Marler, G. D., 1964 (reprint 1978), *Studies of geysers and hot springs along the Firehole River, Yellowstone National Park, Wyoming:* Yellowstone Library and Museum Association, 54 pages.

Marler, G. D., 1969 (revised 1974), *The story of Old Faithful:* Yellowstone Library and Museum Association, 48 pages. These pamphlets are for sale in most shops in Yellowstone and provide good descriptions of the geysers between Madison Junction and Old Faithful.

Muffler, L. J. P., White, D. E. and Truesdell, A. H., 1971, *Hydrothermal explosion craters in Yellowstone National Park:* Geological Society of America, Bulletin, v. 82, p. 723-40. This paper discusses the origin of the ten hydrothermal explosion craters in Yellowstone National Park.

Parsons, W. H., 1974, "Volcanic rocks of the Absaroka-Yellowstone Region": *Rock Mechanics: The American Northwest*, Experiment Station College of Earth and Mineral Sciences, The Pennsylvania State University, University Park, pages 94-101. This is an excellent summary of both the Eocene and the Quaternary volcanic rocks in the Yellowstone area.

Parsons, W. H., 1978, *Field Guide: Middle Rockies and Yellowstone:* Kendall/Hunt Publishing Co., Dubuque, 233 pages. This is one of the better non-technical discussions of the geology of northwestern Wyoming, southwestern Montana, and eastern Idaho. The book contains numerous field guides. It covers a larger area than I have presented in this book.

Pierce, K. L., 1979, *History and dynamics of glaciation in the northern Yellowstone National Park area:* U.S. Geological Survey, Professional Paper 729F. The most comprehensive treatment of glaciation of the Yellowstone Country. This is required reading for anyone really interested in the fascinating glacial history of the area.

Pierce, W. G., 1957, *Heart Mountain and South Fork detachment thrusts of Wyoming:* American Association of Petroleum Geologists Bulletin, v. 41, p. 591-626.

Pierce, W. G., 1980, *The Heart Mountain break-away fault, northwestern Wyoming:* Geological Society of America Bulletin, Part I, v. 91, p. 272-81. In these and numerous other papers referenced in the bibliographies of the above works, Dr. Pierce pieces together the fascinating story of the Heart Mountain thrust that in the vicinity of Cody places 300 million-year-old rock on top of 55 million-year-old sediments.

Reid, S. G. and Foote, D. J. (editors), 1982, *Geology of Yellowstone Park area:* Wyoming Geological Association 33rd Annual Field Conference Guidebook, 387 pages. This book contains a collection of research papers on nearly every aspect of the Yellowstone Country geology by numerous authors. This is probably the best place to start for additional information on the detailed geology of the area. Detailed roadlogs for the professional geologist are included for many roads.

Retallack, G., 1981, "Reinterpretation of the depositional environment of the Yellowstone fossil forest: Comment": *Geology*, v. 9, p. 52-53. Discusses a fossil soil at the base of the three tall petrified trees on Specimen Ridge.

Rinehart, J. S., 1976, *A guide to geyser gazing:* HyperDynamics, Santa Fe, 64 pages. A general non-technical guide that describes how geysers operate and how to watch the geysers in Yellowstone.

Smedes, H. W. and Prostka, H. J., 1971, *Stragigraphic framework of the Absaroka Volcanic Supergroup in the Yellowstone National Park region:* U.S. Geological Survey, Professional Paper 729C, 33 pages. The paper describes the various lithologies of the formations that make up the Eocene volcaniclastic rocks in the Yellowstone Country.

Smith, R. B. and Braile, L. W., 1984, "Crustal structure and evolution of an explosive silicic volcanic system at Yellowstone National Park": *Explosive Volcanism: Inception, Evolution, and Hazards*, p. 96-109, National Academy Press, Washington, D.C., 176 pages. This paper discusses the volcanic system in Yellowstone and presents siesmic data for interpreting the subsurface geology.

Smith, R. B. and Christiansen, R. L., 1980, "Yellowstone Park as a window on the earth's interior": *Scientific American*, v. 242, p. 104-17. This paper provides an easy-to-understand and fascinating discussion of the central Yellowstone Plateau with its associated thermal features.

Tarbuck, E.J. and Lutgens, F.K., 1984, *The Earth: An introduction to Physical Geology:* Merrill Publishing Co.; Columbus, 594 pages. This is a well-illustrated text on general geology. The pictures and illustrations are so good you may want to leave this one out on the coffee table for all to see.

Yuretich, R.F., 1984, "Yellowstone fossil forests: New evidence for burial in place": *Geology*, v. 12, p. 159-162. In this article the author presents good evidence that many of the large stumps on Specimen Ridge are preserved where they grew, rather than transported.

Index

Check for our books at your local bookstore. Most stores will be happy to order any which they do not stock. We encourage you to patronize your local bookstore. Or order directly from us, either by mail, using the enclosed order form or our toll-free number, 1-800-234-5308, and putting your order on your Mastercard or Visa charge card. We will gladly send you a complete catalog upon request.

Some other geology titles of interest:

____ROADSIDE GEOLOGY OF ALASKA	12.95
____ROADSIDE GEOLOGY OF ARIZONA	14.00
____ROADSIDE GEOLOGY OF COLORADO	14.00
____ROADSIDE GEOLOGY OF IDAHO	15.00
____ROADSIDE GEOLOGY OF MONTANA	14.95
____ROADSIDE GEOLOGY OF NEW MEXICO	11.95
____ROADSIDE GEOLOGY OF NEW YORK	14.00
____ROADSIDE GEOLOGY OF NORTHERN CALIFORNIA	12.00
____ROADSIDE GEOLOGY OF OREGON	14.00
____ROADSIDE GEOLOGY OF PENNSYLVANIA	12.95
____ROADSIDE GEOLOGY OF TEXAS	15.95
____ROADSIDE GEOLOGY OF UTAH	14.00
____ROADSIDE GEOLOGY OF VERMONT & NEW HAMPSHIRE	10.00
____ROADSIDE GEOLOGY OF VIRGINIA	12.00
____ROADSIDE GEOLOGY OF WASHINGTON	14.00
____ROADSIDE GEOLOGY OF WYOMING	11.95
____ROADSIDE GEOLOGY OF THE YELLOWSTONE COUNTRY	10.00
____AGENTS OF CHAOS	12.95
____IMPRINTS OF TIME: THE ART OF GEOLOGY	9.95

Please include $2.00 per order to cover postage and handling.

Please send the books marked above. I enclosed $_____

Name_____

Address_____

City _____ State _____ Zip _____

☐ Payment Enclosed (check or money order in U.S. funds)

Bill my: ☐ VISA ☐ MasterCard Expiration Date: _____

Card No. _____

Signature _____

MOUNTAIN PRESS PUBLISHING COMPANY
P.O. Box 2399 • Missoula, MT 59806
✂ Order Toll-Free 1-800-234-5308 ✂
Have your MasterCard or Visa ready.